TO SLIP THE SURLY BONDS...

NASA, the Shuttle Disasters and the Demise of the U.S. Manned Spaceflight Program

**Other Books
by
Jeannette Remak and Joseph Ventolo, Jr.**

Black Lightning: The Legacy of the Lockheed Blackbirds
A-12 Declassified
XB-70 Valkyrie Ride to Valhalla
XB-70 the Return to Valhalla

TO SLIP THE SURLY BONDS...

NASA, the Shuttle Disasters and the Demise of the U.S. Manned Spaceflight Program

Jeannette Remak

SPEAKING VOLUMES, LLC
NAPLES, FLORIDA
2018

To Slip The Surly Bonds...

ISBN 978-1-62815-913-4

This book is dedicated to the three astronauts of Apollo AS-204 fire and the fourteen astronauts that lost their lives in both Challenger and Columbia and all those that worked in the shuttle program and all manned flight programs that the United States has flown. May God Bless them and let us keep them in our thoughts as we look to the stars for better days in the space program of the United States.

Acknowledgments

This book is a labor of love and done out of deep respect for all those in our space program, from the early sub-orbital missions in Mercury to the last shuttle flight. It is most especially, for the Apollo 1 Astronauts, the Challenger Astronauts and the Columbia Astronauts and of course, all of their families.

Their sacrifices have been the backbone of the manned space flight program of the United States. We can only hope that we will prevail in space and maintain our hold in on the space frontier so that their sacrifices will not be in vain.

I would like to thank the following for their help and support: TD Barnes, Joseph Ventolo Jr., Marti Ventolo, Amber Ventolo, The NASA FOIA Office and Ms. Joanne Sibley, Ms. Eve Lyon- NASA- FOIA, the late and beloved Lt. Colonel Donn Byrnes, Lt. Colonel (USAF Ret.) Art Powell, Tony Landis -NASA Photography (Ret), Mary Anne Ruggiero- Phoenix Aviation Research, NASA History Office, NASA Shuttle Program Management, James Petty (The Rocket Man). And last but not least, the one who keeps me sane, my wingman…. my beloved Pekingese…. Shanghai.

If there is anyone that I missed I apologize, I do it without malice and blame my lack of memory. As I get older, these things do happen.

Introduction – NASA and the need to survive

The death of fourteen astronauts* and two shuttles was not calculated consciously, but to be sure they were the result of political and financial manipulation. To this we must add the dereliction of management for the shuttle program. It also heralded the end of the manned space program as we knew it. That means we no longer are sending our Astros into space via our rockets. We are now asking the Russians to do it while we come up with a new plan. The ride with the Russians will end in 2019. We came up with that plan, but it was canceled as soon as one president left office and a new one came in.

The shuttle disasters will forever play in the minds of all who witnessed them. The cold, clear blue sky on that January 28, 1986 morning and the 73-second climb as Challenger and her human crew rose to eternity, is seared into the minds of millions who watched the launch live and on TV around the world. The wait on Saturday morning, February 1, 2003, while the agonizing, deadly silence brought home the horrible reality that Columbia and seven crewmembers had flown home to their Creator. A cold dark reality pervaded not only the U.S. space program and NASA, but the hearts of people all around the world. These are moments never to be forgotten.

There has been much written about how and why it happened, but has there ever been a direct, clear, and concise answer? A comparison of both disasters and a look at the management system that broke and broke yet again, shows us why the U.S. finally ended up with no manned spaceflight system at all. Without a doubt one of the most monumentally stupid decisions ever devised by a president, who had no care for the United States prestige or position in the world, President Barack Obama, made it clear he would rather see a "commercial effort replace" NASA.

Any commercial effort for a man rated vehicle is at least 5 to 8 if not more years away from producing anything capable enough to fly our astronauts to space and return them safely to earth and that is just for low earth orbit to reach the International Space Station. While progress has been made, we aren't any closer just yet.

Commercial space flight is of the future and that is the problem: the future. You can't jump into the commercial realm and expect the aerospace industry, that has been decimated by layoffs and funding cuts, loss of research and development and talent via attrition or just plain lack of will to stay in an industry abrogated by the country's then poor economic outlook, to create "the" spacecraft in a matter of months. In the meantime, the United States is paying Russia $70 million an astronaut to shuttle them to and from the International Space Station, that will be closing down in 2019 as funding runs out. The forty year old Soyuz capsule has been the technology that supported those hard landings for our astronauts. The United States bought and paid for the Russian part of the ISS, thanks to former NASA Administrator Dan Goldin. Goldin's use of funds from the Space Shuttle safety requirements program to support the Russian part of the ISS goes down as one of the reasons the shuttles flew in the condition they did. It was one of the reasons for the shuttle disasters. There were many other problems within NASA and causes found within the Congress itself that added to that scenario. The need to pull apart this technological *Tower Of Babel* and find out what are the reasons for the loss of 14 astronauts and two shuttles and the manned space program is necessary. We need to learn from Challenger, Columbia, the ISS and all concerned in the United States Space Program, what were those negative forces that led to this downfall. Hopefully, we will continue our quest for the new worlds of the universe. There has been some wonderful work done by some commercial companies and leaders like Elon Musk and his company Space X, who were not afraid to pick up the gauntlet and try to come up with a solution. However, that will not happen if we lose our courage to correct and revamp our space program and not kick it to the curb, as has been done by the past Obama Administration. The new Trump administration has signed a Directive to launch us back to the Moon and Mars. We can only hope that there is time enough to meet the challenge and enough want from Congress to reestablish the United States lead in Space.

The question is why was the decimation in the space program allowed to happen? Why did we not refocus on the problems that confronted the Shuttle program from the start, like the foam that fell off the external tank from the first of the shuttle flights and the O-ring that was forever in doubt? Why did NASA, both production and management not correct the issues within the shuttle program involved? Why was there fear in one of the NASA centers of not being in line with the views of its director? Why was Congress so defensive of its pork barrel programs that were forced on NASA, whether they wanted it or not? We see ourselves looking to the CEV (Crew Exploration Vehicle) space capsule, will that fulfill the needs of a space program? Will it move us further into space or just to the ISS and back? Where is the heavy lift vehicle to launch it?

There are many questions to answer when it comes to the future of NASA and the United States manned spaceflight program. We must look to the past to see how it all started to figure where we will end up in the very near future.

This book is a blood chit for seven astronauts that died aboard Challenger and the seven astronauts of the "Grey Lady" Columbia. May they forever fly to the stars and into God's Loving Hands.

Blood chit is a notice that is carried by the military, usually aircraft personnel, that displays messages aimed at the civilians that ask them to help the service member in case they are shot down.

Table of Contents

Introduction   x

What is NASA's Real Purpose?   1

The Start of the Space Age   13

Back at the White House   24

The Start of the Space Transportation System   49

The NASA Centers   81

The Shuttle Begins   103

The Orbiter Tastes the Air for the First Time   132

Painful Losses   148

Columbia the "Grey Lady"   252

The Manned Spaceflight Program   311

Epilogue   331

Appendices   336

Index   353

Bibliography   356

What is NASA's Real Purpose?

In finding out what NASA was brought to life for, let's take a look at the legislation that built it, the National Space Act of 1958[1]. To break down the Space Act into digestible bits allows us to see the main declaration for NASA's existence. The first on the list, is the expansion of human knowledge in both the atmosphere and space. That means that NASA would be responsible for expanding the human aspect of man in space while supporting aeronautics on the ground and in the air.

The next statement is "NASA was intended to foster improvements in performance, speed, safety in the proficiency of space vehicles and aircraft." This is second on the list of the National Space Act. The statement is clear and profound. However, was NASA given the wherewithal to do just that? Had politics and pork barrel deals destroyed that aspect of the Space Act? The Act defines NASA's path. The development and operation of vehicles that are capable of carrying instruments, equipment, supplies and or living beings (human or animal) through space and a safe return to earth is part of that path. This statement gives NASA the mandate to use human, animal and robotics to explore the universe.

The next statement says the NASA needs to conduct long range studies to discover potential benefits and opportunities in both the aeronautical and space activities. The Act also challenges NASA to maintain the United States as a leader in space science and technology. While NASA had two shuttle disasters, the program ran for thirty years and could have run for ten more. We must not forget that the United States/NASA also put man on the Moon via Mercury, Gemini and Apollo programs.

---

[1] Appendix A- 1958 Space Act

However, the current political policy has not allowed NASA to continue to hold the world's title in space science. Albeit, robotic programs have worked well, while suffering funding cuts, these robotic programs have taken us to Mars. It would have been awesome for the United States to follow those robots to Mars via the Lunar base planned for in the new directive given by the Trump Administration.

The Act goes on to state that NASA should pass on relevant information to the Department of Defense and related agencies. Considering that the U.S. Air Force dug its claws into NASA's shuttle program, and without the aid of the USAF, the shuttle program may not have been, the Department of Defense certainly had a lot to say about what NASA did in the early years and even into the development of the Shuttle program.

The Space Act goes on to direct NASA up to and including the shuttle program to facilitate the cooperation of nations within the peaceful pursuit of space and aeronautical activities. NASA has certainly fulfilled this requirement, working with Russia and aiding other nations who sought to work within the shuttle program, along with training astronauts from many different countries and, of course, the ISS with its thirty- two nation partnership.

The Act directs NASA in the cooperation and coordination with other public agencies to avoid "unnecessary duplication of effort, facilities and equipment." Given the amount of NASA centers, along with the necessary redundancy in various programs such as Apollo and the shuttle program, NASA had become entangled in many unnecessary "redundancies."

The Space Act goes on to describe what NASA should encompass. That means that there should be an administrator appointed by the President of the United States. This Administrator would come from civilian life, and his induction would be agreed on by the Congress. The

Administrator would be under the supervision and direction of the president and be responsible for the exercise of all the powers and discharge all the duties of the Administrator's office, and have authority and control over all personnel and activities of the agency. The deputy administrator would also be appointed from civilian life by the president and Congress and perform all the duties that the Administrator should prescribe. It goes on to state that neither administrator or deputy shall engage in any other business, vocation, employment while serving in the position. To note, James Fletcher, former administrator for NASA during the shuttle program was found to have close ties in the Mormon Church and with the political issues of his home state, Utah. His "panel" oversaw the selection of Thiokol as the company to build the SRB. On this statement alone, we see the first of where the shuttle program went wrong.

NASA's Function

We have discussed the National Space Act of 1958 in relation to NASA. Now we can look at the functions of NASA[2]. Simplified they are:

• *To plan direct and conduct aeronautical and space activities.*

• *Arrange for participation by the scientific community in planning scientific measurements and observation to be made through use of aeronautical and space vehicles and to conduct or arrange for the use of scientific measurements and observations.*

---

[2]Appendix B-- Functions of NASA

• *To provide for the widest practicable and appropriate dissemination of information concerning its activities and results.*

• *In the performance of its functions the administrator is authorized to promulgate, issue, rescind and amend rules and regulations governing the manner of its operations and the exercise of the powers vested in it by law.*

These are some of the recommendations that were applied the function of NASA. NASA could also acquire (by purchase, lease, condemnation or otherwise) construct, improve repair, maintain labs, research and testing sites and facilities for aeronautical and space vehicle quarters. It also allows NASA to lease to others such real and personal properties including patents and rights in accordance with the Federal Property Act of 1949. It allows NASA to accept unconditional gifts or donations of services, money or properties, mixed or real, tangible or intangible. There are many more, mundane administrative and legal functions that would be tedious to list. The ones above are the main ideas for our purposes.

There is really no provision to assign priority of purpose to any real tasks. Basically that means it is all the same priority and along with this, there does seem to be very little for the private sector in all the high language outlining the Act and Functions of NASA. What NASA has been given, is a very broad view of how it is supposed to work and what it needs to accomplish but there is no real sense of precedence to what NASA's plan of action should be. We do see, however, that it was easy for the administration of NASA to ease its personal beliefs and wants into the program. It was also easy for many of the political factor to sneak in a pork barrel appropriation when possible. While NASA has its so called orders, just what the mission is comes to this: advance and communicate scientific knowledge and understanding of earth and space to others in the scientific community.

Since NASA as an agency has to compete with other agencies like the Department of Housing, Veteran Affairs for example, for funding, it leaves NASA as a victim of both political parties or any other ambitious policy maker. This means anything which happens, like the Apollo AS-204 fire or the Challenger/Columbia disasters and even international dealings, makes life even more complicated as to how NASA stays afloat. This has left NASA as an agency with others meddling in its decision making processes, not to mention its funding.

A Bird's eye view of NASA

On October 1 1958, NASA officially came into being. This was after President Eisenhower had it sit in limbo while ARDC (Air Research and Development Command, headquartered at Andrews AFB, Maryland) made decisions for aerospace research and the new agency. ARDC was the established go-between for the Eisenhower Administration and the newly formed NASA for at least 18 months until NASA finally came to be. It was also the USAF's little piece of heaven, literally. It did give the military some control over what Eisenhower called the "Space for Peace Policy". Eisenhower was looking for a way to keep the concept of "Peace in Space". He didn't feel he could do this with the USAF running the space program. Next, what was the purpose for the change to NASA from the very successful NACA agency? There had been so much criticism hurled at NASA for the past 40 years or so, but how much of it was truly warranted. Even in the early years there was always some controversy roiling around NASA aimed at its priories, management, and errors that had a common point of origin. In clear language, has there ever really been a specific statement of what NASA's real purpose in life is? If you look past the public relations and the media, the answer is no. There are many in NASA who can give you a very detailed report of every nuance of whatever it is that this particular person touches. This is called compartmentalization. While in some aspects, the "need to know" works well for an outfit like the CIA, it is detrimental to an organization like NASA. An organization like NASA is supposed to be working towards one goal which means sharing all information pertinent to that goal. However, it isn't what NASA has been doing. In the meantime, NASA has been at the whim of every president, Congress and political influence since its inception. It allowed things to be arranged for certain parts of a program that others didn't know about. Hence the first level of the "Tower of Babel.[3]"

What happens now?

Presidents and Congress change every few years. NASA has also had its share of changes. As administrations changed, so did the priority of many projects. A good example is President Jimmy Carter's synthetic fuel, which was cancelled just as soon as President Ronald Reagan took over the Oval Office. Internal and external policy conflicts thrived at NASA. There were so many things competing with each other and so many definitions of statements that were not only left unresolved, but took on a meaning only NASA understood. Early on, it was also President Eisenhower's way of keeping the military happy while conducting his "space for peace" policy. It kept the military industrial complex at bay.

---

[3] The Tower of Babel shows mankind's pride and with his desire to reach heaven and become their own gods. It presents the unification of all people in error. Since pride is one of the Cardinal sins, in essence the sin which condemned Lucifer to Hell, God punished and divided Babel's builders by making them all speak a different language, hence no one understood anyone else.

Missions and change

There have been many questions raised on how the NASA mission changed over time. Government officials, who were defining the space policy at some given time, may see a research and development issue that should be managed by scientists as a matter of national security and defense. Should this mission be handled by the private sector? What did the U.S. policy makers see as NASA's space goals? There were internal wars that developed because of changing issue definitions. There were insiders driving those changes, which caused both internal and external political and economic changes in the perception of NASA's rather amorphous mission.

What all this means to NASA

With the start of the 1950s and Russia's advances in the technology and getting into space, the United States was faced with a huge political question. There was one thing that did happen for the better. The United States Space Program was aided along by various political and personal ambitions within the country. This helped in placing the image and structure of the U. S. space policy within the creation of NASA. This set of broad goals would help the agency early on, but it would prove very difficult to hold onto much later in the agency's struggles during the 1980s.

The development of NASA in the 1950s and its role in U.S. space policy did treat the USSR Sputnik launch as the turning point and galvanized the U.S. into action, which of course resulted in Mercury, Gemini and the Apollo programs. It was the start of NASA, as we know it today. When you look at the agency in its early days, it's not hard to see just what happened here and how a "Tower of Babel" could occur. First, NASA comes into being because the President at the time, Eisenhower, decided that he wanted to see space used only for peace. At the time, the military, the Air Force, was the owner of the skies and anything above it. Because of the Soviet breakthrough into orbital space with Sputnik, Eisenhower was even more afraid that things could go wrong and the United States would be embroiled in a military space race. Eisenhower looked at options and decided that the only way to keep things from escalating and to also to hold onto the United States technological prestige was to change the way things were done. Eisenhower had to figure out who would control and direct the United States in the space program that was coming to pass.

These new activities would be by far, larger and more diverse than anyone could have imagined. Between 1957 and 1961, Eisenhower started to get the answers he needed. Relying on his scientific advisor, James Killian and the RAND Corporation (a think tank developed by Donald Douglas of Douglas Aircraft Corp. 1946) he began to look at the programs that the USAF and the CIA were working on. Those programs, WS 117L the reconnaissance satellite program, AQUATONE the U-2 program, later the follow-on OXCART the A-12 program for the CIA, and the USAF's various hypersonic glide programs were all in the mix. During this chaotic time, administration officials were trying to resist all the domestic political pressure that was coming their way to make changes to the "Space for Peace" policy.

Eisenhower also had to consider what the international community had in the way of its space policy. It was true that the United States did not have a very vigorous space program in 1957. That was due to Eisenhower's very strict fiscal policies. He had insisted that for every space program out there, he wanted to see the results and what it was all worth. He wanted to keep the military out of space for the sake of foreign peace. As far back as 1946, both the Navy and USAF programs were dropped. Project ORBITER was a U.S. spacecraft program, which was canceled in 1955. This was the competitor to Project Vanguard.

The Project ORBITER program was run by both Army and Navy concurrently. The program was later dumped by the "Committee on Special Capabilities", which decided that Project VANGUARD fit the bill instead. Even though the project was canceled on August 3, 1955, the design was used for the Juno I rocket which helped launch Explorer 1, the first satellite launched by the U.S. The VANGUARD program officially began March 17, 1958. Vanguard 1 became the second artificial satellite successfully placed in Earth orbit by the U.S. VANGUARD was the first solar-powered satellite. This project was also held back some 2-3 years. The Soviets called VANGUARD "The grapefruit satellite." due to its small size.

Jealousies and the true start of NASA

There was also the issue that the other military services didn't want the USAF in the lead as far as any space programs went. Inter-service rivalry was rampant at the time. Congress was into the act, stating that much of the population of American taxpayers wanted the appointment of a space "Czar". However, that would not solve any of the aforementioned problems. By adding more layers of bureaucracy to existing programs, it would only slow down any development. Streamlining was called for on the existing Federal and Department of Defense procurement system to help space policy management within the administration to sort out its issues. From February 1957 to Sept 1958, ARDC (Air Research and Development Command, headquartered at Andrew AFB, Maryland) served as the nation's space agency until NASA could get started on its own. Here we can see where confusion, favoritism and basic mistrust developed between the services. The USAF was happy to work within its own confines of ARDC. The rest of the military watched, most notably the Navy, which had its own plans for space that were crushed by the USAF's "ownership of the skies and beyond".

Consider this, a fledging agency NASA, already has a military organization ARDC working while it begins to function and still wants to be considered as a civilian operation. Yet, as a civilian agency, it would exceed ARDC's authority, which is its base. However, it would have to wait eighteen months before it was on its own. That is not to mention that the CIA, at the time, had its own ideas about what it wanted to do. CORONA, AQUATONE, OXCART, were all successful CIA programs yet, what did NASA have to do with any of it? By February 28, 1958, ARPA (Advanced Research Projects Agency) had already canceled WS117L satellite program and authorized the DISCOVERER satellite program. Meanwhile, the USAF was trying to maintain its status quo by holding on to its space programs. Only a few weeks after ADRC's activation, the USAF Chief of Staff, General Thomas White started yet a new space study initiative, saying that he did not want the USAF leadership to be technologically blindsided when it came to the space age, as it was by the missile age of 1947-1954. The USAF was trying to cover all its bases to make sure it wouldn't be undercut, as it had been before, hence the reason for its internal study.

Meanwhile, NASA was still trying to discover just what it was and what it was supposed to do. NASA was indeed on wobbly legs. There was so much infighting by the USAF and other agencies to hold onto their piece of the space pie, very little was going to be left to NASA.

When Eisenhower asked Congress to create NASA on April 2, 1958, he committed himself to a unified national space agency. He wanted it to conduct all space activities except those primarily associated with military requirements.

NASA, the agency, signed into law on July 29, 1958, accepted its very broad mandate to develop the peaceful pursuit of new knowledge and technology in space. Shortly, after all of this, Eisenhower submitted his DOD Reorganization Act to Congress. Eisenhower believed that separate ground, sea and air warfare no longer existed and the services were put on notice about it.

Next to enter the fray was the National Security Council which put together a Board and an informal committee to establish a policy on space and national security. The result was the *U.S. Policy on Space NSC5814/1*. This plan highlighted the significance of the military space program. It placed political implications of operations above the USAF's perceived need for offensive military space systems. Eisenhower's "Space for Peace" policy wanted a passive, automated, negotiation as a means to gain vital intelligence and to ensure international peace. Meanwhile, the USAF put in an "eyes only" study on space and expressed warnings against cutting off the nation's military space program. This study was done after reviewing international space law to insure the nation's security and to assess its effects on military space programs. It would take some five months to complete and would reorient the USAF's approach to military space programs. This wasn't to say that the CIA was sitting back and relaxing. The CIA was very busy working on the picking of contractors for the CORONA Satellite by holding a clandestine meeting at the Flamingo Motel in San Mateo, California in March of 1958.

The purpose of showing these issues is to explain that the development of NASA as a viable space agency, for all intents and purpose, was nothing more than a means to an end. Eisenhower wanted a peaceful space policy. He didn't want the military involved. However, the military had their own ideas about what they wanted to hold onto. Thrown into the mix is the start of ARPA a military agency in itself. CIA was also developing projects on their own. At what point in time was NASA to really take over the space programs for the United States? NASA already was being ostracized by the military who were afraid of losing their own programs.

The CIA didn't want to divulge anything they were working on. What did NASA as an agency really have to stand on? In fact, nothing much more than what NACA (National Advisory Committee on Aeronautics) had to give it. Even that was subject to jealousy and mistrust. We find that the base support for NASA, as an agency, was rife with very unfriendly, mistrusted and covert dealings with other agencies as well as a broad mandate that truly expressed nothing more than a political want, peaceful space.

The Start of the Space Age

NASA, in theory, came into being at the end of the Second World War. The War changed the lives of all the nations that were involved, that meaning the entire world. Before the war, there was not much in the line of space initiatives for any nation. However, there were those few scientists and dreamers that did consider it a possible reality. While there were the dreamers and writers of the 1800s that tried the experiments and wrote their stories, there wasn't much in the line real live proof. In 1857, Konstantin Eduardovich Tsiolkovsky was the Russian scientist who developed the earliest of astronautic hypothesis. Later on, the German scientist Hermann Oberth, and of course, the United States own father of rocketry, Robert H. Goddard, added into birth of the theory of rockets.

However, the real start of the space age manifested itself in the middle of the World War II with the hideous birth of the V-2 rocket created by German engineering. It was the first time that any nation actually took an interest in the idea of rocketry as a possible weapon. It did not look like the thought of traveling to the stars was of interest to anyone in the Peenemunde rocket site or anywhere else in the Nazi regime for that matter, except for a handful of German scientists. In early 1943, all the Nazis wanted was the destruction of England and world domination. Some research has found that the production and development of the V-2 rockets were equal to the cost of 24,000 German fighter aircraft. The United States knew the only thing the Germans had in mind was blistering the British and cost was not something they were worried about. However, at the war's end, the U.S. decided, along with the Russians that they were going to pick the cream of the crop of German scientists in a little project called "Operation Paperclip", which was a basic smash and grab of all the German scientists left over from the war. Russia and the

U.S. were grabbing at all they could find. Werner Von Braun, a Peenemunde brain, found himself in the United States, along with some of his fellow scientists. They had one job: to build a rocket for the United States.

This is a little side note about the U.S.'s decision to take Von Braun. In October 1943, Von Braun had been working for the Germans of course, but he got himself in a bit of trouble when he voiced his opinions. Von Braun made a statement to a couple of his colleagues that the war was not going well and why shouldn't he be thinking about rockets for space instead of rockets for the Reich. It didn't go over very well with the Gestapo and in March 1944, he found himself locked up for the effort. He was incarcerated in a cell at Gestapo headquarters in Stettin, Poland. Von Braun's close associate Captain Walter Dornberger, head of the army artillery and rocket research program, went to Reich Minister Albert Speer to help Von Braun get out of jail. When finally released Von Braun and his cronies found themselves returning to work on a larger version of the V-2, the A-4 which was almost completed in 1944. This A-4 was one in the line of development of the planned manned rocket planes, which would have the ability to make a controlled descent and a manned or unmanned upper stage for a larger missile that had transatlantic range.

Consider how far ahead the Germans were in the role of rocket development. In a closely related program that was called A-6, which overlapped another program of a similar name, Von Braun was playing his cards close to the vest to keep it out of the hands of the German chain of command. This A-6 was a reusable high performance, photo-reconnaissance vehicle, which was launched much the same way the A-4 was, vertically as any rocket of the time. Hopefully reaching the speed of Mach 4, the A-6 would begin its descent to 100,000 feet at which point the ramjet would be ignited. This would let the aircraft cruise to about

1800 mph before returning to its base where it would glide in on a tricycle landing gear at about 100mph. This information is included here to show not only the potential of what Germany had up its sleeve, it shows how in 1944 Von Braun was way ahead of the pack. In fact, looking at the concept of the A-6, it sounds almost equal to the U.S. A-12 Blackbird of the 1960s. This was one of the reasons that while it was known that Von Braun was working with slave labor at Peenemunde and Nordhausen, not to mention working for the Third Reich, all was forgiven in U.S. eyes. The United States desperately needed him to support the U.S. rockets programs. Had Von Braun gone to the Russians, the outcome of the Cold War might have been very different. It is a chilling thought.

Sputnik

The Eisenhower administration which was concentrating on saving money and not allowing the "Military Industrial Complex" to take over its administration, found itself in the midst of the Sputnik hysteria with nowhere to turn.

Sputnik was a product of U.S.S.R/Soviet ingenuity: it was a 184 lb. ball of wires and tubes that made its debut in the vacuum of space, launched by a Russian rocket. While the U.S.S.R. had made it known in announcements before the launch, the United States and the Eisenhower administration didn't pay much heed to it. That is not to say that the United States was hiding its head in the sand, it was just that the United States was busy trying to build a viable missile and create a process for reconnaissance satellites.

There were a couple of hitches in the program, however. The Eisenhower Administration ran on the principal of making sure every penny counted. Eisenhower saw his administration as simplistic when it came to the issue of space and rockets; he wanted it small and quiet. Yet, Eisenhower was insistent on the fact that no military industrial complex was going to taint his administration; his theory was the smaller the size of military spending, the better. He was not going to be pushed into the corner by reactionary generals who wanted every new military toy out there. That unfortunately, was one of the many reasons Eisenhower was left high and dry when Sputnik showed up. While Eisenhower was already into the idea of spy satellites and working on that project, Eisenhower had been warned in advance about the oncoming Soviet leap into space.

When Sputnik beeped its little heart out over the U.S. skies, not only was the U.S. population on the verge of fear, Eisenhower was seriously left out of the picture. It was then necessary for Eisenhower to keep the U.S. citizenry from panic.

It was also at this point that Eisenhower realized that he needed something more than NACA to help with the new issues confronting him. The year was 1957 and it was a very nervous time for the Eisenhower administration. Let us look at what type of space policy the United States had before Sputnik.

NACA

The beginning of the NASA story actually starts with the Eisenhower Administration and the National Advisory Council on Aeronautics, NACA. NACA was a small government agency that was comprised of mostly aeronautical engineers and military test pilots. NACA worked to serve the aeronautical community in the most generic way. It wasn't until later on, as the aviation community reached for speed, that NACA really sunk its teeth into the concept of high-speed flight.

NACA did help to speed the victory over Germany with its contributions to the war effort and aeronautical research, which gave the US aircraft the speed, reliability, endurance and range that won the war.

Perhaps the most outstanding improvement that NACA provided was "drag cleanup", which translates into resistance to airflow. NACA woke up one morning to find itself becoming a new agency. For 25 years, NACA supported the aeronautical community and became the premier aeronautical research firm. When Germany had decided on its campaign to rule the world in WWII, NACA was there, supporting the U.S. in the war effort. Because of the monumental work done by NACA, they earned the title of "midwife at the birth of better American airplanes." The NACA directive to "find practical solutions," gave it importance in aiding the development of aircraft for the U.S. This process provided the aircraft designers and builders of WWII, by use of the Full Scale Wind Tunnel tests, to measure the drag resistance and help the airplane builders on how to correct the problem of drag resistance.

Beginning of NASA

When Sputnik and the U.S.S.R. made its way into space, the creation of NASA wasn't yet on the way. Many of the Congressional proponents for higher science education had been trying for years to get funding for their program. Sputnik helped to push that over the hill and woke up the rest of Congress. The new National Defense Education Act of 1958 helped to get the funding to do just that. Eisenhower helped establish the President's Scientific Advisory Commission (PSAC). This was accomplished years before it became attached to the military. Eisenhower's move was to place scientific expertise and scientists within the White House itself where he could better control the spending. Eisenhower appointed James Killian as his presidential scientific advisor. Killian, was president of Massachusetts Institute of Technology from 1948 to 1959. Killian was the first of the true scientific presidential advisors. He headed and oversaw PSAC and was therefore instrumental when it came to laying the groundwork for NASA.

On February 8, 1958, Secretary of Defense, Charles Wilson announced that within the Department of Defense, along with Eisenhower's instructions, was he created the Advanced Research Projects Agency (ARPA), which would direct the rocket and space agency, later named DARPA (Defense Advanced Research Projects Agency). DARPA acquired the responsibility for all long range, defense related research and development. Federal spending for research and development increased quite a bit in the wake of the USSR's Sputnik. The National Science Foundation managed to more than double its funds, which rose to 44.4 million in 1956 and up to 102.1 million in FY1958.

Between 1955 and 1960, total federal funding on research and development rose from 2.6 billion to 7.4 billion for non-defense agencies and 9.6 to 23 billion for research and development. Both houses of Congress had set up new committees that were now devoted to space. This would go on as the basis to the creation of NASA. Congressional proponents argued over the primary issue, which was national survival or national security. The development of space capabilities, civilian and military were now at the deepest levels of the government. There were council members, which included the Secretaries of State and Defense as well as the new soon to be NASA administrator. As much as he would have liked to control it even more, Eisenhower was not for this type of approach to space policy. He opposed it because he thought the National Security Council would suck up the entire program and never allow it to turn it into something important.

These objections held up final approval of the Space Act until Eisenhower and then Senator Lyndon Baines Johnson could come up with a compromise that being the president would chair the council, keeping it under presidential control and direction. Congress persisted in trying to catch up to the U.S.S.R. while Eisenhower stated that there was no competition. While signing the Space Act, he made no mention of the

Soviets and only the most indistinct mention of the United States leadership in the space age. At a meeting, Eisenhower finally announced T. Keith Glennan as the new NASA administrator. Glennan did not start out in the aviation/aerospace field. Early on in his career, he actually worked in the motion picture business and then found himself briefly working for Vega Aircraft Company. In 1942, he joined the Columbia University department of War Research where he served as administrator. He then moved on to the Department of the Navy for underwater sound labs. At the end of the war, Glennan became an executive of Ansco Corporation in NY. In October 1950, he joined as a member of the U.S. Atomic Energy Commission and on to the administrator of NASA position. In his speech, he didn't mention the Soviets at all and said that he wanted a sensible space program. Right up to the end of Eisenhower's term, he never once said that there was a "space race".

By 1958, the definition of the Space Policy was established and except for the president himself, administration officials had little choice in dealing with Congress or attempting to reassure the public. They had no choice but to accept it. Glennan would put in his diary, "We are not going to attempt to compete with the Russians on a shot for shot basis in a way to achieve space spectaculars". During Congressional hearings and in public, Glennan made use of a more eloquent statement. At an authorization hearing in 1959, Glennan gave an answer to questions of how to justify expenditures of millions of dollars to explore the unknown. He listed scientific knowledge, applications, creation of new industries and understanding the unknown universe. Yet, he still said there was another overriding reason for the U.S. space program and space exploration: Russian space activities are devoted to activities as a nation furthering communism's design on mankind. NASA's role as a defender of national security was highlighted by the Senate modifications to the Space Act, thus giving NASA a significant role in National Defense.

Rockets, rockets and rockets

When Eisenhower came into office in 1952, he had another concept in mind for rockets. He wanted them to fly in surveillance against United States enemies. Eisenhower created the International Geophysical Year for that purpose and announced it at the Geneva summit in 1955.

The IGY was proposed in hopes that the U.S. and U.S.S.R. would keep from each other throats. Eisenhower's "OPEN SKIES" policy, allowed both the U.S. and U.S.S.R. to peek into each other's backyards to see what was going on and make it an even playing field. The U.S.S.R. did not like the Eisenhower idea at all. However, this gave Eisenhower the ability to approve new secret surveillance program to hide in IGY. The GENETRIX balloon and the U-2 spy plane were used for the purpose of surveillance. There was a study by the RAND Corp. that supported the idea of satellites as a more successful way of getting surveillance. The RAND Corporation was started on May 14, 1948 as an offshoot of World War II. RAND's parent company was Douglas Aircraft. It was then broken off from Douglas to become a nonprofit, independent business venue for research and development. It dedicated itself to the purposes of scientific, educational and charitable studies for the benefit of the welfare and security of the United States. In 1955, when Eisenhower received the report from his science advisor James Killian, he approved the strategic satellite program WS-117L. Eisenhower had brought both space research and the process of intelligence gathering. No longer could it be treated as just a science or weapons program. Space meant something.

Eisenhower went on to set up the IGY (International Geophysical Year) with the idea of bringing scientists of the world together. This also caused suspicion on the part of the U.S.S.R. July 29, 1955, the Eisenhower administration announced that the president had approved plans for a series of small earth satellites to be launched during IGY; this helped Eisenhower hide the new satellite program which meant using any means possible to protect his its secret. Misdirection and diversion was used to divert attention from WS117L. Eisenhower used Werner Von Braun as a visible means of diverting attention from the security of the program.

However, the U.S. lost its first satellite in a launch attempt. Eisenhower was questioned as to why he didn't use Werner Von Braun as the head of the IGY program and not just as a ploy to cover up the satellite program. The selection of the Naval Research Labs as the program manager of the satellite program was a disaster, which was a result of politics and the old standby, in service fighting. Later on, there was a more sophisticated satellite on line with better tracing possibilities. Von Braun wanted to use the Redstone rocket as early as 1956 to launch the new satellite.

In his State of the Union message on January 6, 1956, Eisenhower made note of the increased importance of long-range missiles. $1.275 billion was scheduled for the fiscal budget of 1957 to be used for guided missiles, with an additional $1.43 billion for military research and development. By January 10, 1956, the first U.S. built liquid rocket engine with a thrust of 400,000 pounds was fired for the first time at a California test site. Later on in the month for the USAF, the Northrop *Snark* missile was fired from Cape Canaveral. As time went on, the political line up of different offices and military departments were working for the interchange of technical information on all the missiles in progress.

More and more developmental processes went on to insuring that the missile program of both the civil and military would be combined and that it would establish a more centralized coordination of all the military services on the development of earth satellites.

The change in the research effort was shown by a breakdown on the years of 1955,1957,1959: This chart showed how the interest had shifted in the NACA/NASA issue. The Air Force and NASA were working out of a very small station at Edwards AFB. The high-speed flight prominence was centered on the problems of winged spaceflight, coming from the days of breaking the sound barrier. Walter Williams was the first head of the Dryden Flight Research Center for NASA, and later NASA's chief engineer in the Washington D.C. headquarters. He came up with at least a dozen concepts which would affect the design and flight of future high-speed, hypersonic aircraft. There were many design problems which were covered in aerodynamic heating, heat transfer, aerodynamic efficiency, crew survival, along with some pilot problems, poor landing configuration, large accelerations, reaction control operations. Williams, as head of the High Speed Flight Station for NASA, knew that they had a lot to do with the X-15 rocket plane and the upcoming Mercury program. On September 27, 1959, the station was renamed the NASA Flight Research Center. Once the center became viable, Williams had left for Project Mercury and Paul F. Bikle would now be in charge of the new center. Bikle came in on September 15, 1959 and stayed for twelve years. Bikle's main challenge was to get the X-15 up and running. Bikle loaded on an additional eighty people into the staff to help do just that.

Research	1955	1957	1959
Satellite	5%	11%	16%
Ballistic missile Research	1	1	5
Boost glide a/c	15	18	35
Anti ICBM	1	1	2
Surface to air missile	3	4	4

Advanced fighter	33	32	16
Supersonic bombers & Transport	23	19	18
Subsonic bombers & Transport	18	12	3
Special projects VTOLS etc.	1	1	3

Back at the White House

NASA became embroiled in the problems and controversies that would later endanger astronauts of the Shuttle program. We need to look back at the early days and how the race for space changed from a reason to outdo the U.S.S.R and keeping the Russian wolf from the door, to the complex and bureaucratic nightmare NASA became.

Along with the youth and resilience that the newly elected president John F. Kennedy brought to the White House, there were other issues afoot. The first was the Cold War. Sputnik put the fear of God into Eisenhower and JFK built his presidential campaign for more activism in space.

JFK pledged, if elected, this nation was not only "going to be first, but first in space." Kennedy used the growing power of the "imperial" presidency, gotten in part from the Cold War mindset, to both empower the nation with the responsibility and need to execute a "crash program to place Americans on the Moon." Kennedy would have his problems starting with the Bay of Pigs disaster and then the Cuban Missile Crisis. The addition of an intensive space program could help detract from those issues. Of course, back in the USSR, Kennedy's counterpart Khrushchev was quick to sound off about the triumphs of the Russian space program and trying to roll that into a third world invitation to join the communist party. Kennedy argued that U.S. failures in space weakened U.S. prestige. No truer words were spoken, then or now. However, the U.S. experience with creating the right rockets and boosters was a trying and often disappointing time.

Kennedy knew the Soviet economy was weak. He also knew that the USSR had problems with heavy lift power in their rockets. Meanwhile plans went forward to get the Mercury and Gemini programs up and running at NASA. NASA managed to get the first U.S. chimps into a sub orbital flight. Alan Shepard had the first manned suborbital flight and followed by John Glenn's first orbital flight around the earth. The Mercury program would close out as a complete success.

Kennedy was sure that the moon was the place to go by the end of the decade. Kennedy talked a good game regarding space but his internal problems were getting more pressing. April 20, 1961 found JFK right in the middle of the Bay of Pigs debacle. He had called in then Vice President, Lyndon Johnson, also chairman of the Space Council and asked him to come up with something "first in space". JFK wrote Johnson a memo: "Do we have a chance of beating the Soviets by putting a lab in space or by a trip around the moon or by rocket to land a man on the moon and return? Is there any other space program which promises dramatic results in which we could win?" JFK was desperate for anything to take the country's mind off the ineptitude of his administration. Space was just the way to do it.

Kennedy went on to ask Johnson, "How much additional would it cost? Are we working twenty-four hours a day on the existing program? If not, why not? Are we making maximum effort?" Are we achieving the necessary results? I would appreciate a report on this at its earliest moment." With that memo, JFK laid the weak foundation for using NASA as a tool for prestige and power and not for the investigation of our last frontier. By the next day, Kennedy was playing the space card to the hilt. He told reporters," If we can go to the moon before the Russians, we should." This is the prime political example of the "rock" NASA is built on, political need, loss of faith in his administration and something to help his flagging administration regain credibility. Kennedy had the right idea for space but not the right need. Political expediency is part of the weak link in NASA's foundation.

LBJ and the Space Council

Lyndon Johnson went to work on his president's wishes. Johnson had already used Sputnik to great advantage in the Congress to garner support for space. As vice president, he was merely a shadow in the JFK administration and the JFK limelight was a bit dim at this time. Kennedy had racked up the balls for the break shot in an aggressive effort to move the space program ahead via his vice president. April 22, 1961, NASA's Hugh Dryden, deputy administrator responded to LBJ's letter concerning the moon program by telling him that there was a chance for the U.S. "to be the first on the moon and return safely IF a determined national effort is made." Dryden added that the earliest date would be 1967, but costs could range about $33 billion and that would be $10 billion more than the entire NASA budget for the next ten years.

Johnson turned to Secretary of Defense Robert McNamara. McNamara knew Kennedy well and he knew JFK was moving to a view of space policy. After three months, McNamara already had a handle on the verbiage and a way that would win the president over. Except for Robert Kennedy, the watchdog and Attorney General for the administration, McNamara had already proven himself a leader in the administration. With this knowledge, McNamara knew Kennedy wanted a space program and an aggressive one at that. McNamara had another impetus working behind the scenes.

The increased effort that would make a perfect opportunity for companies in the aerospace industry that were already irate over the cutbacks that McNamara was planning (i.e. the XB-70 Valkyrie bomber USAF program) in the defense budget. It would be a way to get them off his back.

McNamara wrote a memo to Johnson," Major achievements in space contribute to national prestige. This is true even though the scientific, commercial and military value of the undertaking may, by ordinary standards, be marginal or economically unjustified. What the U.S.S.R will do and what they are likely to do, therefore matters of great importance from the viewpoint of national prestige." There you have it, McNamara was going to make sure that he could appease the aerospace industry by throwing them a piece of the space/NASA pie in lieu of a few military programs that he was cutting back.

Johnson searched out friends in private business as well as other government officials including Werner Von Braun, who was now running the Redstone Arsenal in Alabama, which later became the Marshall Space Flight Center. Von Braun told Johnson that the United States had a chance of sending three men around the moon ahead of the Soviets and that there was an excellent chance of beating the Soviets to the first landing of a crew on the moon with return capability, of course. Von Braun added that one significant complication was the need for a jump by a factor of ten, over the present rockets that would be necessary to accomplish this. While today we may not have such a rocket, it was unlikely the Soviets had one at the time. Hence, the United States would have to enter the race towards this next goal in space exploration against odds favoring the Soviets.

The Russians were having issues with their heavy lift rockets and Kennedy knew that there was still a need for an intensified program in the United States. Von Braun felt that the program could be accomplished by 1967. Von Braun concluded with this statement, "I do not believe that we can win this race unless we take at least some measures which thus far have been considered acceptable only in times of national emergency." In short, NASA needed more bricks in a hurry and damn the money spent. Kennedy would at least be able to announce the program would be underway. His assassination in November of 1963 would end his vision of the program. Johnson began to persuade many of the political leaders of the need for a very aggressive program. When Johnson heard of some problems in the Senate on the newly aggressive lunar program, he asked, "Would you rather have us as a second rate nation or do we spend a little money?" The mandate that Kennedy gave Johnson was framed so bluntly that the vice president was unlikely to ever go back to the president and say, "Don't worry about the space race, we have other problems". This could not have been a truer statement as Johnson was working hard to influence both Kennedy and the administration. On April 28, 1961, Johnson gave Kennedy the report he was after. Johnson said, "The U. S. has the greater resources than the U.S.S.R for allowing space leadership but has failed to make the necessary hard decisions and to bring to those sources a way to achieve leadership. This country should be realistic and realize that other nations will tend to align themselves with the country, which they believe will be the world leader--the winner in the end."

Dramatic accomplishments in space are being increasingly identified as a major indicator of world leadership. We are neither making maximum effort nor achieving results necessary if this country is to reach a position of leadership."

Johnson insisted the manned exploration of the moon was essential whether the United States was first or not. Yet, Kennedy still wanted to have a strong reason for backing a presidential initiative to start Project Apollo. However, he had moved closer to a larger consensus for the ways and means of bringing both key government agencies and business together.

Bits and pieces for Apollo

Both Mercury and Gemini programs were moving ahead quickly enough in an attempt to catch up with the Russians. While John Glenn went on to spend five hours circling the earth, it still wasn't quite enough to beat the Russians. As the Mercury program changed out to Gemini, which was a larger spacecraft that could now accommodate two astronauts, along with rendezvous dockings and spacewalks, the ten missions of Gemini, flown from 1964 to 1966, gave the United States the shot in the arm it needed in space. The United States prestige was getting better against its Soviet counterparts and the exemplary records of both programs, Mercury and Gemini, laid the yellow brick road to the moon. However, while Mercury and Gemini were flying high, the next step for the Kennedy agenda was to reach the moon. In 1963, James Webb, NASA administrator, had to find the money to produce a space miracle. He had managed to squeeze $3.5 billion out of Congress to get Apollo started and hold it through the decade.

Remember that Congress was free handed with the money to support the space program up to now, building on the success of Mercury and Gemini. The president's wish to go to the moon was still feeling the money Congress had pumped into those programs. Apollo was still in good stead as far as Congress was concerned, now that Apollo had good funding and was a tangible program. James Webb wanted to expand the "definition" of Project Apollo beyond just the mission to the moon and landing/return. As a result, even those three projects, not officially funded as of yet, were justified for support of the mission. Webb had his reasons to do this.

He and the rest of NASA leadership were committed to design a broad based space exploration program not just a single project even though Apollo was extensive.

Because of the massive size of the lunar landing mission and cost, there was little opportunity to undertake additional large space exploration initiatives. Using Apollo as that vehicle to accomplish many scientific and technical additives was a practical solution to funding problems.

By attaching all these other projects to Apollo, Webb managed to get NASA to incorporate five other groups into the program and helped to get cost support of those involved. The result was more space science.

Research, education and other projects carried out under the auspices of Apollo, might not have happened otherwise, (i.e. moon probes, Surveyor program and Lunar Orbiter projects.) Due to those ideas, Webb managed to cover all those programs under Apollo. As an example, Apollo would need a radar tracking system, and communications systems. NASA justified the development based on the lunar mission more than the landing project. Webb also had a mid level "educational" effort to train aerospace engineers and scientists using Apollo as a "hook". He wanted to use it to expand U.S. education and research by channeling millions into national education institutions via Project Apollo. The center for this was developing and funding of the university program in 1962 in Apollo's name. Hence, by the time 1970 rolled around, NASA had paid the bills for graduate education of more than 5000 scientists and engineers at a cost of over $100 million. It also spent $32 million on construction of university labs and gave more than $50 million worth of multi disciplinary grants to some 50 universities. What this all says is at the beginning of Apollo, funds were given out for all and sundry in the name of Project Apollo. Much of which had no direct course with Apollo or were on the edge of the program. This is another rotten brick in the foundation for NASA. The money was allotted for NASA to use for Apollo, NASA took it and ran with it, putting everything it could under the Apollo umbrella. This really shows the misuse of taxpayer money and trust; even if you wanted to justify it by saying NASA was building its infrastructure of scientist and engineers.

Who's watching the store?

It was obvious that no one was watching the store or bank accounts while NASA used the APOLLO program as a "cash cow" to garner support and make political friends. The taxpayer did benefit somewhat, but not until years later. While "padding" in projects had been done for years, even by the Pentagon to cover black projects, NASA is a civilian agency and APOLLO was not a black program.

At the time this was done, it was James Webb, NASA Administrator, free-handing on NASA's behalf, based on the reasoning that at the time he was afraid that the plug would be pulled on the budget and political support for the project would fade. It eventually did. Webb used the funding umbrella to grow his own benevolent services. This couldn't have been done without the full knowledge of Congress, and its pork barrel mentality. It is just one more instance of bureaucratic corruption.

Briefly, NASA development was an offshoot of the Cold War. The base definition for its internal policy began to change when APOLLO came onto the scene. JFK was under great political pressure to revise the APOLLO program spending due to the Republican Party and its angst on program spending. Many of the Democratic Party was also questioning JFK's spending on space. JFK really wanted space cooperation with the U.S.S.R. to hold off any space race or expansion on the Cold War. All of this did much to push NASA's focus more on the moon landing instead of planning for a solid view of what its true position would be. The bottom line was NASA had been created as a Cold War tool. While Eisenhower didn't feel the need for manned space, Kennedy felt the need to compete and attempt to outdo the U.S.S.R. for political reasons. He did not live to see the results and how different the program may have been if he had survived.

NASA on the other hand, now had to produce the moon landing as a memorial to the slain president. They accepted the money doled out with both hands. Because of the free spending policy of early NASA management, a weak root grew and began to struggle with a top-heavy bureaucratic system. As NASA funding dried up later on in the years following Apollo, so did the ability to manage what successful programs it did have.

Yes, Congress was obstinate and didn't fund the agency for some time. The initial groundwork should have been laid for a successfully run program, without the top-heavy bureaucracy management or wasteful program development spending, but that isn't what happened.

Had the proper steps been taken, NASA may have been able to develop the shuttle more effectively instead of having to compromise the program and the vehicle the shuttle became. This is the core of the shuttle management failure. NASA would never stand a chance of becoming the agency it wanted to be unless it could find a way to get off the treadmill of politics and pork barrel spending and its own bureaucratic level building. NASA needed to be able to spend wisely and work with a budget consciousness that was never part of its plan. Instead, being all things to all people, NASA's identity was driven by the Cold War, a weak presidential administration and fear of being cut off from funding. The building blocks for the ensuing shuttle program were already quite weak and would grow weaker still.

How we got to *DYNASOAR*

On October 1951 through January 1952, NACA was busy with the Bell X-1 aircraft, which was the forerunner of the X-15. In 1952, the Aerodynamic committee of NACA fronted the proposed plan to devote more time to hypersonic studies, yet it was not enough. At the time, NACA was bogged down with many research projects.

On March 9, 1954, NACA headquarters directed the labs to put forward their ideas to Washington D. C. for an appraisal. The Ames Research Lab along with Edwards AFB and the Langley Research Center agreed with the concept. By October 1954, the USAF and NACA along with the Navy created a hypersonic committee to come up with the specifications for the proposed aircraft.

In August of 1953, a proposal was given to Edwards AFB Complex, via NACA headquarters, for a five-phase hypersonic research program, which included and led the way to the basic concept of an orbital winged vehicle. However, the USAF felt this as being too futuristic and dumped this idea. Yet, the need for the piggyback, two stage to orbit research aircraft led to one of the earliest precursors of the shuttle.

By the end of 1953, the idea for hypersonic research aircraft had fostered two military study efforts. The first was from the USAF Scientific Advisory Board and the other from the Office of Naval Research. In the annual meeting of 1954, NACA was given the go ahead to procure a hypersonic research vehicle.

By December of 1954, both the USAF and the Navy agreed to fund the design and development segments of the project and the USAF would control the different segments of the program as it progressed. On the completion of the contractors testing, the aircraft was turned over to NACA to conduct the flight tests and report the results. The "Memo of Understanding" that was written included the phrase "accomplishments of this project is a matter of national urgency."

X-15 Rocket Plane

The USAF, NAVY and NACA, which created the committee, included an interagency group of senior executives (Hugh Dryden did the representation for NACA) for the project. The committee really hadn't exerted much influence on the project which was known as the "X-15 Committee". It served as the honorary committee, which was more show than blow at this stage. It met occasionally, and usually provided a rubber stamp for the project. In addition, it was useful (to get budget approval) to say "And here is what the X-15 Committee wants to do" said a senior engineer with the project. The committee stayed alive until Oct 27, 1957, when it was closed down.

The X-15, funded December of 1954, hit the flight line in 1958, brought with it many historic discoveries. The concept that the hypersonic boundary layer flow was chaotic and not laminar was one huge discovery. There were also further discoveries that chaotic heating rates were lower than predicted by mathematical theory.

The X-15 also brought the knowledge through direct measurement that hypersonic aircraft skin friction heating was lower than predicted. The tests found that hot spots were generated by any surface anomaly. Another facet found that the transition from aerodynamic controls to reaction controls could be used in rockets. They were transitioned back again, and easily blended without incident. The X-15 also brought energy management methods for all reusable launch vehicles following their re-entry from space. The X-15 program single handedly opened the door to the idea of manned space rocketry. Considering the USAF was getting the raw end of the deal by having to close down its hopes for a space program and join with the newly formed, not yet operational NASA (NACA and ARPA were still running the show for space), it was not too pleased about events as they were. The USAF wanted to go into the world of near orbital space and the way that they would do it was with lifting bodies. With *DYNASOAR* cancelled before it ever got operational, the X-15 was its only hope.

The X-15 was a most complex space plane to begin with. The X-15 was in parallel development with the *DYNASOAR* project. Both NASA and the USAF saw the need to acquire hypersonic aerodynamics and heating information. The X-15 came into being to serve the purpose for high-speed research aircraft and really showed its merit. Created by the North American Aviation, the X-15 was a rocket powered space plane that was part of the USAF and NASA research forum. Both NASA and the USAF saw the need to acquire hypersonic, aerodynamic and heating information.

The X-15 was responsible for setting speed and altitude records in the early 1960s. It reached past the very edge of outer space and returned data that was later used in both spacecraft and aircraft designs. During the X-15's program run, it made 199 flights (12 pilots that are listed below) and met all the USAF spaceflight standards exceeding some of the requirements by as much as 50 miles.

Two of the X-15 flights met the criteria for actual space flights. There were eight different X-15 pilots made astronaut in the X-15. Joe Walker flew two astronaut qualifying missions with the X-15. Mike Adams made astronaut on the X-15 flight that killed him.

List of pilots who flew the X-15:
    A. Scott Crossfield, North American Aviation, 14 flights
    B. Joe Walker, NASA, 25 flights
    C. Robert White, (USAF), 16 flights
    D. Forrest Petersen, Navy, 5 flights
    E. John McKay, NASA, 29 flights
    F. Robert Rushworth, USAF, 34 flights
    G. Neil Armstrong, NASA, 7 flights
    H. Joe Engle, USAF, 16 flights
    I. Milton Thompson, NASA, 14 flights
    J. William Knight, USAF, 16 flights
    K. Bill Dana, NASA, 16 flights
    L. Michael Adams, USAF, 7 flights

Lifting Body Research - The Path to the Shuttle
    The heart of the space shuttle started with the research on lifting bodies. *DYNASOAR* was one of the first of the lifting body programs that almost made it to fruition. The original purpose for *DYNASOAR* program was to develop a single-pilot, reusable spacecraft. *DYNASOAR* was first

for this type of manned project that got as far as the development stage. The fact that *DYNASOAR* was cancelled, just before its first drop test from a B-52, was a waste of taxpayer dollars.

DYNASOAR's lack of completion is sad, however the "reinventing the wheel" virus has run through NASA and the U.S. Space Program since the inception of the agency. Since it originated from the "cost conscious "do everything in the house, NACA", of all the concepts that NASA did not inherit from it predecessor was "do it once and do it right". NASA tossed away a great deal of possibilities and once again, the U.S. Space Program would find itself not only reinventing the wheel but also attempting to do it with the lowest bidder.

DYNASOAR was developed as a manned- winged, space plane that would use a rocket to boost it into hypersonic speed and low earth orbit. The program had three different profiles:

DYNASOAR I: was termed a hypersonic research vehicle that was intended for a low earth orbit. Speeds would increase as the test program expanded. *DYNASOAR I* used a second stage booster that incorporated liquid fluorine/hydrazine along with a Bell Corp. engine called Chariot. Time was an issue. If the Bell product was not ready on time, the use of the single Atlas engine, or the X-15's XLR-99 engine, would be used instead. In March 1963, this plan called for an airdrop of *DYNASOAR I* followed by a single stage, sub-orbital booster flight. By the end of 1965, with two stages, it would expand to orbital flights.

The Characteristics for DYNASOAR:
Crew: One pilot
Length: 35 feet 4 inches
Wingspan: 20 feet 10 inches
Height: 8 feet. 6 inches
Wing area: 345 feet 2 inches

Empty weight: 10, 395 lbs

Max. take off weight: 11,387 lbs

Max. speed: 17, 500 mph

Range: earth orbit: 22,000 nautical miles

Service ceiling: 530,000 feet.

Rate of climb: 100,000 feet/min

Wing loading: 33lb/feet 2 inches

DYNASOAR II: In its next phase, *DYNASOAR II* was planned as a hypersonic reconnaissance vehicle. Using the *DYNASOAR I* assembly and it would follow the same parameters of altitude and speed for *DYNASOAR I;* it increased by the use of a second stage. The pilot would control the reconnaissance systems. The reconnaissance system would consist of high-resolution camera, a side looking radar and ELINT (electromagnetic radiations from foreign sources could be picked up and used as electronic intelligence).

The Atlas booster used along with a Bell Chariot engine (if the other assigned engine was not available) would be ready for an airdrop by a B-52 Stratofortress, in January 1966, followed by booster tests in 1969. A weapons system was added in 1969.

The Lockheed A-12 Blackbird, which was already in service by 1962, and flew until 1968, was supposed to replace this version of the DYNASOAR II that never came to be.

DYNASOAR III: was considered the ultimate hypersonic bombard-ment and multi stage launch vehicle to reach orbital velocity. January 1970 was set for the first flight tests, followed by the orbital flight in mid 1971.

DYNASOAR III's Boost Glide was more favorable than the advanced turbojet and ramjet engines of the day like the J-93 General Electric engine of the XB-70 Valkyrie.

A rocket boost glider would have a speed range of Mach 5 to Mach 25, needed by the mission. Of course, any air breathing engine was always more complicated and expensive to develop. It would also work at lower Mach speed. Anything flying below Mach 9 according to the think tank RAND Corp. could be vulnerable to the 1965 brand of Soviet Air Defense.

There were scheduling and funding issues during those early years of the *DYNASOAR* program, just as there are now in today's political world. *DYNASOAR* was a needed asset because the USAF was worried by the fact that the ICBMs did not have sufficient stationary target accuracy.

USAF and its ideas

The USAF had its own ideas about lifting body theory. NASA's success with both the X-24A and B strengthened the USAF intent to continue with its lifting body story. NASA and the USAF were joined at the hip when it came to the new M2-F1 lifting body aircraft.

In 1962, at the Dryden Center in California, NASA developed a program for lifting body prototypes. The M2-F1 was designed as a lightweight, unpowered prototype aircraft to flight-test the wingless lifting body concept. It did look akin to a "flying bathtub" but was constructed only of plywood and tubular steel. The frame looks like something from a children's playground monkey bar. It used many off the shelf parts including some from Cessna. The M stood for manned while the F stood for flight. The construction was finished in 1963. The construction was done in partnership between NASA Dryden and a company by the name of Briegleb Glider Corporation. With a very meager budget of $130,000, (1964 dollars) NASA engineers and Briegleb engineers got to work.

The first flight test was executed on the Rogers Dry Lake Bed in California. The M2-F1 aircraft was towed by a hot rod Pontiac car, pulling it across the lakebed at about 120 mph. This test proved the M2-F1 had sufficient flight worthiness to continue with more tests. Later on, a C-47 "Gooney Bird", the workhorse of the military services was used to aero-tow the M2-F1. August 16, 1965, the M2 –F1 now had a newly added ejection seat installed and small rockets.

The C-47 towed the M2-F1 to 10,000 ft at approximately 120 miles per hour with the M2-F1 trailing on a 1000 ft. towline. As the aircraft was released, it flew back to the lakebed. The M2-F1 program proved the lifting body concept. The reality of the program is that it was done at such a small cost. There were 77 flights. The M2-F1 success story led to NASA's development of two more lifting bodies of much heavier weight.

M2-HL-1

NASA and the USAF were fused together when it came to the M2-F1 and the new HL-10 (Horizontal Lander). A new program was coming on line called PILOT (Piloted Low Speed tests), which was within the larger SV-5 *START (Spacecraft Technology and Advanced Reentry Test Program)*. This was a USAF contract for the development of a heavy weight lifting body design for manned flight. This was the center of the START program. It was felt to be superior to NASA's M2-HL 10 lifting bodies because of a better lift to drag ratio and a better cross range ability. It was also more advanced in aerodynamic efficiency. This really showed the USAF's determination in lifting bodies research.

At the heart of the USAF work was the Flight Dynamics Laboratory (FDL) at Wright-Patterson AFB, Dayton, Ohio, established March 8 1963. It was re-organized from the Directorate of Aero-Mechanics. The FDL had an interest in hypersonic vehicle design. They examined a variety of lifting bodies, winged bodies and other concepts. The majority of these designs ranged from low lift to drag (L/D) to high lift to drag (this is the amount of lift generated by the shape of a wing, divided by the drag it creates when flying through the air).

The USAF was very enthusiastic about the high lift to drag concept and managed to develop a whole family of FDL (Flight Dynamics Lab) shapes that would lead to a high lift to drag hypersonic performance. Some of the early designs showed promise in solving issues of some lifting reentry and orbital flight issues. Studies, expanded into a series of incorporated variable geometry configurations, showed where the vehicles could enter the upper atmosphere as a lifting body and then change to a variable sweep wing to increase lift to drag ratio, during an approach and landing. This was of course a revolutionary approach to the entire idea of lifting bodies. To add a variable sweep wing to a fixed fuselage (like the wings of the B-1 bomber) brought forth the idea of "interference" configuration could use complex under surface designs which would position the shock flow for better lift much like the XB-70 Valkyrie used compression lift. To improve the supersonic lift to drag, with the aid of down folding wing tips that also improved the stability of the aircraft at high speeds. The USAF's Flight Dynamics Laboratory (FDL) however, did not restrict its studies to just lifting bodies: it actually included major concepts that were used in the development of the space shuttle. The X-24A and its heirs, which we will discuss shortly, remain as the most discernable concepts that the USAF devised in lifting reentry development, following the DYNASOAR program.

The PILOT program (Piloted Low Speed Tests) was a manned transonic and supersonic test vehicle, which was known as the SV-5P of the START program. By the end of December 1964, PILOT also forced a great deal of key management issues such as the number of vehicles that were needed and who would do the actually building of the vehicles. It seemed that Northrop Corp. had the upper hand; because of the experience, they had building the M2 F1 and the HL-10. January 26, 1965—The USAF held a conference regarding the PILOT program. Many things needed to be decided:

- Should it be a single rocket powered vehicle
- Should it consist of a rocket powered and jet powered vehicle
- Should it consist of a single jet powered vehicle

There were other things to be considered during this USAF conference such as cost, schedule, and resources such as the amount of B-52s support available and flight simulators. The current leaders of the program, Colonel Curtis Scoville who was in charge of PILOT briefing, General OJ (Ozzie) Ritland (of A-12 SR-71 Blackbird fame) and General E.B. Giller in USAF headquarters continued into another meeting on February 15, General Ritland's February 24th letter to the USAF headquarters regarding the START program in which he recommended that the single vehicle, low speed program be expanded with the addition of a second rocket and one jet powered vehicle. AFSC (Air Force Systems Command) contracted for all three vehicles. Since the Martin Marietta Corp., on a sole source basis, would build the single jet powered vehicle, a third B-52 was added to the two already on hand at AFFTC (Air Force Flight Test Command). The total cost would be $8.9 million through 1968. The value of the second rocket-powered vehicle would soon be a back up in the event of a loss or damaged vehicle. While the cost of the project was deemed viable, the difficulties imposed by the 1966 funding which was extremely tight, which was a problem.

The X-24A- NASA and USAF's Best Hope

The Martin Marietta Corp. rolled out the X-24A, July 11, 1967 with full ceremonies at the Middle River Md. plant. After spending the fall of

1967 in flight test, the first scheduled flight for the X-24A would be in early 1968. Testing was conducted at Edwards AFB in California. In October of 1967, NASA and the USAF signed a *"Memo of Understanding"* for the support of the X-24A. This memo confirmed the earlier NASA-USAF connection with prior lifting bodies programs.

While the X-24A was not the physically sleek beauty like some of the other lifting bodies in earlier research, it still was the ticket to the show. Its shape was different from the HL-10 and the M2. The difference in the design was X-24A had wings that turned up into an almost 90 degree angle with the fuselage, while the M2 HL-10 had short, stubby wings at a wider angle to the fuselage. The X-24A would later meld into the much sleeker X-24B. Martin Marietta shipped the aircraft out to the NASA Ames Research labs for subsonic, full-scale wind tunnel tests.

When Ames was finished, the little aircraft was on its way to Edwards AFB. Flight test would start in early 1969. On April 17, 1969, the maiden flight, a glide flight completed successfully of the X-24A was made by pilot Jerauld Gentry. Gentry later flew the X-24A on its first powered flight on March 9, 1970 where he reached into the transonic regime at Mach 0.87. After this flight, Gentry and NASA pilot John Manke along with USAF test pilot Major Cecil Powell piloted the X-24A through its performance envelope with full success.

On Oct 14, 1970, John Manke flew the X-24A on its own initial cruise past Mach 1 and reached Mach 1.9 at 67,913 feet. Approximately two weeks later, Manke again flew the X-24A to 71,000 feet, which is what the space shuttle would be designed to do years later. In a landing approach, June 4, 1971, the last test flight, was a little of an anticlimax due to a malfunction in two of the XLR-11 engines. The aircraft could only make subsonic speed. While the X-24A was small, it gave much in the line of research. Its one hang up had to do with the rocket engine shutting down prematurely. However, pilot Gentry made a safe emergency landing. There was damage to four of the flaps used for maneuvering and some wiring was burned.

The X-24A had one small quirk during boost; it produced a wicked nose up trim change that halted any chance of a low angle of attack during testing in powered flight. The engineers figured out that the aerodynamic effects of the rocket plume exhaust, caused a nose up condition. This item later led the shuttle designers to look out for similar problems.

Beside that issue the X-24A (which was built by Martin Marietta and known also as the SV-5P) led to one significant accomplishment: the SV-5 shape was the only one evaluated in actual free flight along with hypersonic, supersonic, transonic and subsonic velocities.

Like the M2 and HL-10, the X-24A, which was the fourth of the lifting bodies series to fly, showed that shuttle type hypersonic vehicles could make accurate landings without engine power. The X-24A was able to show it could land on a dry lakebed and NASA engineers were not worried about landing on concrete runways at all.

The lifting bodies tests were building great confidence on the fact that an unpowered shuttle could land on any conventional runway after its return from orbit. It was time to move onto the X-24B.

X-24B

The X-24B would be the last of the rocket propelled research aircraft. The Flight Dynamic Laboratory (FDL) in the latter part of the 1960s brought about three different re-entry shapes: the FDL 5, 6, 7. All of these shapes were rated for hypersonic flight which could reach Mach 4 onto to orbital velocities at Mach 8 thru 12.

The USAF was hoping that the shapes would be able to be used for sustained hypersonic cruise aircraft using air breathing engines as well as a boost glide, orbital reentry vehicle with the ability to land on any conventional runway. While FDL saw the chance to go further with the new shapes, it wanted to use the same configurations used with the M2 HL-10. They wanted to build a low-speed, manned, demonstrator vehicle for transonic, subsonic and supersonic testing. The next plan of action for the FDL was to experiment with the new shapes by "gloving" them onto the SV-5 jet powered training aircraft *(the SV-5J was the jet powered version of the SV-5)*, which the Martin Marietta Corp. was looking to build for the Test Pilot School as trainers. However, those trainers were cancelled for safety reasons.

The "gloved" F-7 FDL shape was wrapped around the SV-5/X-24B body, which was modified to incorporate the three vertical fins of the current SV-5 configuration. It was called the FDL-8. In January of 1969, the FDL had a new development plan and that was to drop the FDL-8 from a B-52. While all these studies began to take shape, it showed how jet propulsion could be an advantage. The new "gloved" configuration was tested on the X-24A. Based on wind tunnel tests done at Wright–Patterson AFB in Ohio, it showed the new FDL–8 shape had really great potential. The FDL idea moved ahead. However, USAF Major General Paul T. Cooper, chief of Research and Technology balked at the X-24A "gloved" concept. He opposed it to the point of bringing it before the Joint Air Force Scientific Advisory Board. The Air Force Flight Test Committee briefed the USAF Scientific Advisory Board and the panel agreed the USAF could not do without the project.

By the end of August 1970, NASA, FRC (Flight Research Center Edwards AFB) and the AFFTC felt that while the program was worth the time, effort and money, Air Force System Command delayed the approval while agreements were worked out with NASA-USAF joint funding. On March 11, 1971, NASA passed over $550,000 to the USAF for work to begin on the "gloved" aircraft. By June 24, 1971, the X-24A (developed from the X-23 PRIME program) had completed its final flight, albeit with problems. The new version of the X-24B was about to come online. January 1, 1972 found that the USAF had awarded the Martin Marietta Corp. the contract for the new "gloved" configuration. This followed with Grant Hansey, Assistant Secretary of the USAF for Research and Development and John Foster, Director of Defense for Research and Engineering signing off on the joint *"Memo of Understanding"* between NASA and the USAF concerning the new X-24B program. The X-24B was now the official new program with a price tag of $1.1 million for the new research vehicle. If built without "gloving" the aircraft would have cost over $5 million to develop.

More testing at the USAF Arnold Engineering Development Center found the FDL-8 shape doing well at hypersonic speed. However, the question remained what would happen when it slowed down to subsonic speeds. In the fall of 1972, Martin Marietta delivered a new X-24B:

- It was now longer in length from the X-24A by fourteen and a half feet.

- Weighed in at 13,800 lbs.

- There was a new double delta planform allowing for better center of gravity control.

- A boat tail to help with subsonic lift to drag.

- Flat bottom

- Three-degree nose ramp, which allowed for hypersonic trim.

The X-24B used "off the shelf" components which was a great aid in cost controlling. The XLR-11 rocket engine was also included in the new X-24B. Back at Edwards AFB, where flight tests began, pilot John

Manke made the first flight in the X-24B on August 1, 1973. This consisted of a glide flight launched from a B-52 at 40,000 feet and flew the descent landing 460 miles per hour. Manke experimented with maneuvering at 200 mph finally landed on the lakebed. The first powered flight for the X-24B occurred on November 15, 1973, which as also made by John Manke.

October 24, 1974, brought the fourteenth test flight for the X-24B, which reached Mach 1.76 (1164 mph) which by far was the fastest speed attained by the aircraft. March 22, 1975 found the X-24B reached its highest altitude of 97,000 ft. before landing on the lakebed. The handling qualities of the X-24B were pleasing to all the test pilots. Subsonic handling also seemed to be a great asset and the aircraft got super marks in all aspects

The question remained, however, could the X-24B make an approach and landing similar to that required by a space shuttle. The space shuttle design phase was already under way and all these questions were part of the X-24B's job to find out. Could the X-24B make a landing on a concrete runway? By January 1974, the proposal to find out was under way and approved by the X-24B Research Sub-Committee. On August 15, 1975, John Manke again launched from the B-52, lit the XLR-11 rocket engines and touched down at the 5,000-foot mark on the Edwards concrete runway. Manke said, "We know now that concrete runway landings are operationally feasible and that touch down accuracies of 500 feet can be expected. We learned that the concrete runway with its distance markers and unique geographical features provides additional "how goes it" information not available on our current lakebed runways." September 9, 1975, saw the last of the X-24B flights.

It would also be the last of the U.S. post WWII rocket research programs for aircraft. The X-24B following the last research flight by NASA pilot Bill Dana, soon to be a shuttle astronaut and Captain Francis R.

Scobee (late of Challenger) along with NASA pilots Einear Enerablson and Tom McMurty completed six familiarization flights in the X-24B. November 26, 1975, the X-24B dropped from a B-52 for the last time on its 36th flight, piloted by NASA's Tom McMurty. The final word was "All objectives completed." The X-24B was retired to Edwards AFB where it later was sent to the National Museum of the U.S. Air Force at Wright-Patterson AFB, to sit next to PRIME along with a mock up of the SV-5J (jet powered version produced by Martin Marietta), which was made up to look like the X-24A. It was the crowning glory of a success-ful hypersonic program. It was not all over just yet, however. The Flight Dynamics Laboratory (FDL) had one more idea. They wanted a hyper-sonic aircraft that could make the Mach 5.5 level using the XLR-99 engines from the X-15. The proposal became the X-24C. It was a combi-nation of the NASA hypersonic research aircraft, and rolled into yet another program called NHFRF or the National Hypersonic Flight Research Facility. Unfortunately, while many tried, the acronym came out to be "NERF", not quite a prestigious name for a program.

X-24C

The X-24C became the most analyzed and researched "non-flown" hypersonic vehicle. Everything was done on paper and in the house, so to speak, as far as the FDL was concerned. While the Air Force Systems Command was all for this unique program, the hedge point was it was going to cost money, almost $60 million. The X-24C had its friends within the NASA crowd, however there were still many other hypersonic programs going on at NASA's Langley Research Center.

With all the work, NASA and the USAF had done during the 1960s and 1970s, and the huge demand on resources for the X-15, things were financially tight. In July of 1974, with high program cost of the imminent concern to both agencies, the requirements for an advanced, air breathing hypersonic vehicle had changed.

By December of 1975, both NASA and the USAF decided to form the X-24C Joint Study Committee. December 10, 1975 brought a *"Memo of Understanding"* between USAF and NASA, however the NHFRF really had no real hold due to the complexity and high cost. In July of 1976, the NHFRF became a joint NASA-USAF program, which would be a test bed for hypersonic research. However, the USAF had its ideas for a hypersonic vehicle and NASA had its ideas. This is where the X-24C morphed into the NHFRF. The X-24C look was gone and with it went the way of the NASA concept. Due to the cost and budget conflicts within both agencies and NASA's commitment to the Space Shuttle, the program was cut in September of 1977. The NHFRF program was over.

There is some fantasy to this story. Northrop had the M2, Boeing had DYNASOAR (called the X-20) proposal that never happened, the mini shuttle proposed by the Flight Research Committee of the USAF and of course the X-24C and the NHFRF. They are all projects or concepts that might have been. Hypersonic research and the aircraft created for the testing of it would be the base the shuttle was built on. The M2 HL 10, DYNASOAR, X-15 and the X-24B added so much to the history and creation of hypersonic research and the core the shuttle would eventually be built upon.

The Start of the Space Transportation System

The Space Shuttle went through many metamorphoses before it came to be the vehicle we now know. It took the program many years of development and many agencies to bring it to fruition. That perhaps might be the main problem with NASA. NASA unlike NACA was not self-sufficient, it never was and could never hope to be. The Shuttle was the product of these agencies starting with the Department of Defense, the USAF, the many internal panels and boards of various agencies within the government and Congress and various aerospace companies that all had something to say about how to design and build it. There were many programs that preceded the shuttle, like the Dynasoar and other lifting bodies that were discussed in earlier chapters. What NASA finally called the Space Transportation System or the Space Shuttle was the product of the lifting bodies research.

The USAF and the Shuttle

The USAF study requirement SR-89774 was used to research recoverable space boosters. In 1959, the ROLs (Recoverable Orbital Launch system) was on the drawing board. It had some great concepts, like a single stage design, which could take off horizontally, and use an air collection system that would distill, compress and liquefy oxygen that mixed with liquid hydrogen to feed the engines. This was called the LACES (liquid air collection engine system) It was a unique and forward thinking program. Another system, ACES (Air collection and enrichment system) along with some other scramjet concepts were out there for design study. However, there were huge technical problems, which required aerial refueling at Mach 6, something we haven't accomplished yet in 2018, except for the X-54 Waverider from DARPA. HIRES

(Hypersonic inflight refueling system) was almost fed to the X-15 program but it was stopped before anything catastrophic could happen. None of these very technical and highly risky programs had much in the line of supporting advocates; hence, the programs were killed off.

You must admit, the USAF was working hard to move the study of hypersonics ahead not to mention engine technology. Yet, they didn't have the right mix, at least not yet. The USAF had also invested in the TSTO (Two Stage to Orbit) vehicle and the SSTO (Single Stage to Orbit) vehicle. These two programs were in the mainstream right around 1962. The USAF had shifted its plans to the TSTO early in the new aerospace studies program. Some designs wanted a large payload bay, and others wanted a smaller payload bay, yet the aerospace vehicle and propulsion panel of the Scientific USAF Advisory Board said that such a requirement was "premature because the program was still in very early stages." There were many who liked the plan because it gave a great deal of flexibility and it was a reusable launch system for the military missions which would include orbital, strike, supply and rescue along with a cheaper way of getting payloads into space other than with an expendable rocket. The seed for the reusable orbital vehicle was planted. The concepts and the orbital strike mission later appeared in the shuttle discussions.

The Aerospace Industry

The giants of the aerospace industry and the early space program, Lockheed, General Dynamics, McDonnell Douglas, Republic Aircraft, Goodyear and North American Aviation (later called North American Rockwell) decided to undertake systems designs for an aerospace plane.

General Dynamics, McDonnell Douglas and North American Aviation received a $500,000 contract from the USAF Aeronautical Systems on June 21, 1963 for planning studies. Martin Marietta was already in a contract to the USAF Flight Dynamics Lab (subdivision Structure) to build a full-scale wing fuselage that would represent the space plane. *The USAF Study Requirement 651* gave the aerospace industries many useful research studies on air collection enrichment systems, subsonic combustion ramjets, scramjets, advanced turbo ram jets also known as turbo accelerators, structural materials and aerodynamics. However, time was already running out for the Aerospace plane or the *651 Study*. Many were not happy with the time wasted on the concept. The Air Force Scientific Advisory Board made a statement in early 1960, that it had *"warned it was gravely concerned that too much emphasis may be placed on the more glamorous aspects of the Aerospace plane, resulting in neglect of what appears to be more conventional problems."* The USAF did not want to go any further with this concept. By October 1963, the Aerospace plane was finished.

The USAF Scientific Advisory Board lost all faith in the program and said no more work would be done on it. It all came down to the biggest problem the USAF had and that was trying to identify an aerospace program that actually came out of a requirement for a fully recoverable space launcher, which at that point in time didn't have the hardware or design ability to support it. The USAF may have been ahead of its time, but it sure wasted a lot of money.

It was clear the USAF had to focus on clearing the way in the technical fields and make the path solid for when they could project payloads into orbit. This had to happen every year to increase the need for recoverable launchers and make it a competitive source. The USAF Scientific Advisory Board section of the aerospace vehicles and propulsion board felt there was such a chaotic history attached to the space plane along with so many indefinable issues.

The Aerospace Plane and the USAF

With the fact that that there was so much ridicule concentrated on the space plane, the board decided to *"forever drop the name"* Aerospace Plane. It also wanted the USAF to increase their vigilance so that no new programs *"would achieve such a difficult position"*. Congress drove the final nail in the coffin for the Aerospace plane in FY 1964 budget. The Department of Defense refused to go to bat for it. The work that was done and money spent, the USAF tossed the program out due to the fact it didn't like the publicity focused on the name *"Aerospace Plane"*.

The Department of Defense decided that the USAF had to reorient the *651 Study Program* to address the needs of hypersonic cruise flight within the atmosphere. USAF Systems Command agreed, but argued that the hypersonic cruise studies should pay attention to the potential of such systems and launches of space payloads. With the newly modified *651 Requirement* to study sustained hypersonic cruise atmospheric vehicles, hypersonic vehicles that worked the upper atmosphere for launch platforms, manned and unmanned second stage space payloads, we still need to remember that the *Aerospace plane* while having a disappointing history, still gave much to research.

On the hunt for the shuttle

While the *Aerospace Plane* was an exhaustive attempt to define a technology and mission requirement for a shuttle system, it also allowed for the SSTO and the TSTO approach to space flight. It did give many design and research studies that did influence the latter shuttle designs. What is interesting is that it didn't matter if you were looking at NASA or the USAF, the degree that civilian industry relied on the government to furnish those concepts, instead of designing their own, was odd. The government basically furnished a shape and the aerospace industry added the contents by *"filling in the blanks"*.

The *Aerospace Plane* concept was a single stage to orbit vehicle, which truly brought science fiction to life. With the technology of the 1950s and 1960s and no ability to adapt to the concept, it did bring with it the promise that future systems would see the aerospace plane alive in some form. Both *Dynasoar* and the Aerospace Plane represented a unique example of military interest in space by using a lifting body vehicle. However, they weren't the only ones. In the early 1960s, the studies by both industry and government on the potential for lifting vehicles for military and commercial use were out there. The many studies that came out of the *Aerospace Plane* showed how both the aerospace industry and the military saw the potential in lifting body vehicles, and their studies would bring forth the shuttle later on.

However, there was no real incentive on the part of the aerospace industry at that time. It was more of a "paper napkin" casual design; handed to the aerospace company and the company took it from there to develop.

Where was the initial research and development? The Aerospace industry was dancing to the tune of the government and military instead of the other way around. That is why even today as we struggle to find a commercial successor to the shuttle and have to wait for our astronauts to return from the space station via 1960s Russian technology, the aerospace industry today is stuck in the mire, trying to design its way out. This begs the question of; Did the U.S. and the Obama administration not only abandon the manned space program but sabotage the commercial chances of developing a system to put the U.S. back into a manned space program on its own, by asking the commercial aerospace industry to develop what it doesn't have the wherewithal to do, at least not yet? Progress has been made recently but it will take years before there is a tangible, safe system to return U.S. astronauts to space via U.S. hardware. What happens in the meantime? We pay the Russians to lose our astronauts communication on reentry, bounce them all over the earth once they land and pay them $70 million. This program will be coming to an end financially in 2019. Not only was the Obama administration at fault for not having a back stop measure to protect the U.S. position in space; the Obama Administration did not support the initial start up of the commercial aerospace industry to pick up the ball left from the shuttle close out which was started by the Bush Administration. Yet as soon as Obama entered office, he canceled the "Constellation" program, signed on to by George Bush that would have started us back to the moon at very least. Obama felt that the program was a non-starter as there was no Congressional interest in it and no place in his budget for it. Hence, the lighting quick end of Constellation, thank you, President Obama.

NASA and the contractors

While all the discussion above may seem unnecessary in determining how the shuttle was built from the early 1960s on down to later years, each case of dealing with the USAF or NASA designs it was imperative that the contractor tow the line. Since the 1950s-1960s, the government was dictating exactly what the commercial sector of aerospace contractors would build. It was not until the shuttle program became a reality that the freer hand was given to the contractors. With the development of the FDL (Flight Dynamics Lab) of the USAF, all the lifting body shapes for Lockheed's shuttle type vehicle and the FDL worked closely together. The studies done on the high Lift to Drag reentry vehicles starting with the FDL's fixed geometry F3 and variable geometry V-4 design to the FDL/5 shape, were part of the Lockheed study report of 1969 written in response to a USAF contract. It consisted of the stage and a half concept or a space transportation concept for a space transportation system.

The Lockheed Delta, lifting body orbiter with a small variable sweep wing, which was between two huge fuel tanks, gave the configuration its name "Star Clipper." This orbiter had two features of FDL work. First, the stage and a half concept with parallel tanks. This came from an FDL paper done by the American Institute of Aeronautics (AIAA) written in May 1968 by Alfred Draper and Charles Cosenza. This idea was to use two jettisonable external tanks, and the second design was a high lift to drag FDL reentry shape that developed into the Lockheed vehicle call the *"MX Configuration."*

There is something very interesting in this concept. The amount of similarities between the abilities sought in the MX vehicle and the ultimate shuttle, which reduced operating and development costs. A partial toss away configuration, a large payload bay and high payload weight, made the argument that partially expendable designs like the stage and a half concept could make larger orbiters more attractive, because they could carry propellants outside and allowed more room for internal payload. The development of this vehicle was big enough to carry their propulsion, fuel and payload as a fully reusable structure. The paper would become important later on in the early 1970s, as the NASA shuttle began to change from a huge two stage fully reusable vehicle with fly back boosters to a smaller more effective and less expensive semi-reusable system.

Lockheed went on to submit its "Star Clipper" stage and half design to the USAF and NASA for evaluation to show how a commercial "generic "configuration had to meet the needs of many agencies. The basic design wasn't accepted because NASA didn't want to acquire a system that was dependent on external tanks or orbiter aerodynamic integration abilities. The "Star Clipper' worked on these abilities and ironically they later showed up on the shuttle. NASA agreed on the technical feasibility of the "Star Clipper" by using the parallel tanks and the "V" concept devised by another commercial vendor McDonnell Douglas, and looked at by NASA's study group for the Shuttle. The paper developed by Draper and Cosenza was similar in concept.

Reports and more Reports

With the end of the *Aerospace Plane*, the USAF received its marching orders to redirect its studies to hypersonic vehicles. It all led to the study of the hypersonic launch platform. The FDL started its studies of the RLV and RSLV, known as the *Reusable Launch Vehicle* and the *Reusable Space Launch Vehicle*. The USAF had continued interest in the lifting body reentry systems, which could work in space and still interest NASA as a successor to Apollo program. This led to the joint Department of Defense/NASA Aeronautics and Astronautics Board and the ACCB (*Aircraft Change Control Board*) report. This board was to establish an Ad Hoc panel on reusable launch vehicles, which consisted of NASA and DOD joint chairmen and a DOD representative along with some ten representatives from NASA.

The committee was officially established August 24, 1965 yet closed just about a year later. It produced a report in September of 1966. This panel examined the many reusable launch vehicles and the making use of hypersonic air breathing engines and rocket powered stages. The panel considered the numerous cost uncertainties. The technical risks needed resolution and encouraged development in manned earth orbit activities. Yet, the ACCB's panel could not identify a single concept capable of satisfying the needs for NASA and DOD. The panel summarized a variety of proposed systems, which included using a horizontal and vertical take off, single versus multistage configurations, air launching and air breathing engines and rocket propulsion. The panel also felt that a number of key technical areas were inadequate, including a major portion of the aerodynamics and propulsion fields.

The ACCB panel was uncertain about how much was going on. However, it was optimistic that lifting reentry vehicles would be a reality as early as 1974. Their report showed how the ACCB panel was limited in its views. Even those experts who were on the shortsighted side when it came to seeing and predicting many of the technological problems of lifting reentry vehicles were hesitant. The panel had recommended partially reusable launch vehicles with the hope the cost would spread out over a shorter time for flights thus making it competitive with the throwaway systems of 1975.

The throwaway systems were a dream as far as the shuttle that was built was concerned. However, it would still dominate the scene until NASA was forced to reverse its concepts after the budget axe hit. The ACCB panel saw their potential ideas for vehicles:

- Class I—Ready by 1964-75
- Class II—Ready by 1978
- Class III—Ready by 1980.

The differences in the classes were based on how sophisticated they were, and it took until 1981 when the North American Rockwell Company came up with the raw shuttle design.

NASA PHASE A: The Shuttle from the Start

NASA, much like the late 1950s and early 1960s had put many people, panels, agencies and such into studying what was to be the ultimate shuttle. The Marshall Space Flight Center had worked closely with both the government and industry along with the USAF, which in the 1960s sponsored shuttle like studies from private industry.

All the major aerospace firms were involved, Lockheed, Boeing, McDonnell Douglas, along with Convair, Martin and North American Rockwell. In June 1964, the NASA Committee of Hypersonic Lifting Vehicles had agreed to development of a two stage to orbit shuttle. While NASA was involved with ACCB panel, it was already thinking about its own shuttle concept. It is true that NASA really hadn't thought much about what its next step would be after the last APOLLO mission was flown. When it finally did focus on what its next move would be, NASA thought about the orbiting space station or maybe a trip to Mars. Mars was far away in NASA's technological future so it based itself on the space station which also meant it would need some type of vehicle to fly there. NASA had two ideas:

• A higher, faster, design was coming from the hypersonic airplane to the spacecraft as she finally developed into.
• The Shuttle needed to travel from earth to the orbiting space station and back.

In between these two concepts was the shuttle. However, to be honest, there was no great momentum for building it like there was for the Apollo project.

According to LeRoy Day, NASA deputy director of the Shuttle Program, his first exposure to the Shuttle program came when his boss Dr. George Mueller had called him into a meeting with some fellow engineers in early 1969. While Day did arrive late, he heard what Mueller had to say regarding the shuttle. There was nothing in that meeting that Day had expected, namely Apollo, which was in full swing. Mueller asked Day would he support the shuttle program. Day answered, he could probably do something right after the Apollo XI flight readiness review, thinking it was a short-term project.

It was then that Mueller made it clear to Day that he wanted him to get involved with the shuttle, then and now, to quote *"tomorrow morning"*. Day was shocked but got to work and established a Shuttle Task Group (STG).

Under his direction, he would evaluate both what the agency needed and all the system concepts that went with building a space transportation system. On Oct 20, 1967, the NASA Marshall Space Flight Center in Huntsville Alabama, and the Johnson Space Flight Center in Houston, Texas, issued a joint request for a proposal for an eight-month study of an 'Integral Launch Reentry Vehicle System (ILRV). This proposal would not result in an actual development program. The study was looking for economy and safety rather than optimized payload performance. This was the start of the "Phase A NASA Shuttle Studies."

- PHASE A consisted of advanced studies.
- PHASE B Project Definition.
- PHASE C Design
- PHASE D Development /Operations.

This all changed over time with PHASE A becoming the preliminary analysis, rather than advanced studies. Regardless of the ACCB Panel's uncertainties, the panel was excited over the possibility of lifting body re-entry launch vehicles coming to be, with the hope that this lifting body re-entry technology would be available by 1974 for Class II and 1984 for Class III.

The wish for both air breathing propulsion and scramjets also entered into the panel's hopes. The idea behind all this was to help competition within the development process, which would bring down the number of competitors so that NASA would restrict the PHASE C design stage to only firms who would compete in PHASE D. It was under this system that NASA would develop the shuttle. In February 1969, NASA was finally awarded study contracts to Lockheed, General Dynamics, McDonnell Douglas, and North American Rockwell for the PHASE A studies.

The Leroy Day task group had been evaluating all the shuttle needs and five months after the start of PHASE A award in July 1969, the

SSTG (Space Shuttle Task Group) issued a study report of its efforts which ended with the ILRV class system would have been able to handle six major space missions which would encompass:

- Logistical support
- Orbital placement and retrieval of satellites
- Delivery of propulsion stages and payloads
- Propellant delivery to orbit
- Satellite repair and maintenance
- Short duration manned orbital mission

The SSTG already had a preference for a fully or near fully reusable system which endorsed the shuttle as the "keystone to the success and growth of future space flight development for exploration and beneficial use of near and far space." The problem was in finding a design to satisfy all of it.

Any design that would have to answer all these requirements would definitely have tradeoffs, which could include:

- A partial or fully reusable system
- If winged, either as a delta or straight wing or variable sweep wing.
- Fly back piloted booster
- Off the shelf engines or a new propulsion system
- A winged or lifting body
- A low cross range (200 nautical miles) versus (1100 Nautical miles)
- Small versus a large payload bay and capacity
- Sequential stages of parallel burn stages.

The SSTG divided the concept studies that it reviews into three different classes:

- Class I: recoverable orbiters using expendable boosters
- Class II: Stage and a half concept
- Class III: fully reusable TSTO vehicle (twostage to orbit)

NASA determined that the needs of the space shuttle to carry to the space station components would require a larger payload bay, 15 ft.x 60 ft. and 50,000 lbs of payload. Each of the six configurations consisted of:

Class I: the MURP (manned upper reusable payload) a lifting body based on a NASA configuration which would use a variable geometry wing and use a landing engine. It was dropped because it had poor landing visibility, possible heating problems and a very complicated wing pivot mechanism.

CLASS II: The second and third configurations were the Lockheed Star Clipper and McDonnell Douglas parallel tanks stage and a half concept.

The Triamese remained in **CLASS III**. The fourth would be a **TSTO** design based on NASA's HL10 lifting body shape. Lockheed was thinking of using HL 10 with a booster to launch a second HL10 into space. Both the Orbiter and booster would have air-breathing engines to return to the landing site.

The fifth was by General Dynamics and it was the **TRIAMESE** study. Convair studied the variable geometry lifting reentry vehicle since 1965. The TRIAMESE consisted of three equally sized and aerodynamically similar stages joined to form a two-stage booster with orbital stage. This concept was by Convair and used by the USAF space and missile systems (SAMSO) and with NASA's interest in space transport systems, Convair evolved the concept into an orbital vehicle for NASA. This differed from NASA's concept because it was a marked attempt to reduce costs by using three basically similar vehicles. Recurring costs were considered more important than minimal developmental costs. Convair received serious consideration but didn't make it through **PHASE A**.

The sixth configuration went through widespread analysis and was the basis for many design spin-offs. The Max Faget straight wing design concept consisted of a low wing design of moderate aspect ration having a blunt nose, slab side fuselage.

More noteworthy was the straight wing, which gave the spacecraft the look of a 1940s airliner. Faget (resident guru of NASA and NACA) liked small payload shuttles, having a minimal cross range of 200 nautical miles, which meant his design would have a steep angle of attack when it entered the atmosphere. This design was the core of subsequent NASA shuttle studies, until the agency's engineers were forcibly made to recognize that the vehicle possessed several fatal flaws, such as a sus-pected inability to withstand the thermal environment of reentry.

This was a major consideration in any design for a re-entry vehicle. January 1971 had come and NASA shifted its studies to the delta plan-form and increasingly concentrated on a partially reusable system con-sisting of an orbiter with external propulsion and propellants. The Faget design remained as a basic contender. What is even more remarkable is that it remained a viable concept in some NASA divisions even after the evolution of the 040 baseline orbiter (a delta shape that directly evolved into the shuttle).

During the PHASE A studies, North American Rockwell and McDonnell Douglas looked at the Faget concept. Rockwell had a TSTO design, which used Faget's concept for both the orbiter and the booster. McDonnell Douglas used the HL10 shape as the orbiter and Lockheed had its "Star Clipper".

General Dynamics held onto its TRIAMESE design. Martin Marietta submitted a totally unsolicited study for its own complex "Space Master" which was a lifting body between parallel boosters joined by a stubby wing and tail surface. The joined boosters functioned as a single vehicle; something like the layout of an old F-82 twin Mustang with the ability to come back to earth. Out of all these, North American Rockwell won the prize.

Rockwell and the Shuttle

The USAF and NASA joined up for studies of reusable aerospace vehicle formation as part of the President's Space Task Group. Vice President, Spiro T. Agnew chaired the Task Group, formed by President Nixon on February 13, 1969. This led to closer bonds between USAF and NASA as it involved senior level participation now. In early 1969, the FDL was aware of Faget's design and dismissed it in two in-house reports in June and November of 1969. These reports pointed out the problems with reentry without experiencing structural failure of the wings by aerodynamics effects of testing.

The FDL followed up with two AIAA[4] papers, one that represented NASA in Oct 1970 and one in Jan 1971 by which time the shuttle was in PHASE B Studies. When you look at this miasma of studies, plans, phases, holding onto concepts that were known to be useless, and on it goes, how much time and money was really wasted in bureaucratic redundancy? If the shuttle was already in PHASE B, why was FDL *(Flight Dynamics Lab)* still looking at Faget's design and calling it unusable when that had already been established? The FDL already knew that the delta shaped wing was the way to go and so did NASA. However, some NASA advocates still felt that the Faget approach was the way to go and that the USAF did not appreciate the expertise that NASA had in Lift to Drag approaches with a high angle of attack, a critical point that could not be overcome in the Faget design for operating

---

[4] American Institute of Aeronautics and Astronautics

safely. Even after a visit from some NASA engineers to Wright Patterson AFB in Dayton, Ohio, the home of FDL, the feeling persisted within NASA's spacecraft design division in Houston that the USAF was not taking them seriously. When the Faget concept finally did die out, as much from internal NASA dissension, as from external criticism by USAF specialists, NASA was still skeptical about favoring a lifting body approach.

This does show some of the problems infecting NASA, in that the ACCB panel (Aircraft/Airframe Configuration Change Control Board) initially was naïve enough to think all this could happen in such a short time. This was, by all means, not only tunnel vision, but the basis for many of the future issues that NASA faced. The panel was recommending reusable launch vehicles technology and thinking it would cost less to develop than a fully reusable system. Costs could be spread out over a shorter time with less flights, allowing for the spacecraft to be competitive with its throwaway systems was really out of dreamland. When the shuttle was finally built, the basic detail made sense. However, the dream of a fully reusable system would "dominate" shuttle thought for almost another five years until the late 1970s.

The STS (Shuttle Transportation System)

In 1970, the USAF Secretary, Robert Seaman, along with NASA Administrator, Thomas O. Paine established a joint NASA/USAF Shuttle Coordination Board. The STS committee, which was co-chaired by Grant Hansen, assistant secretary of the USAF for Research and Development, along with Dale Myers, who succeeded George Mueller in 1969. Dale Myers was a former North American Rockwell executive who now was the associate administrator for NASA. Myers had a strong background in aircraft design and development stemming from his time at North American Aviation. At this time Michael Yarmovich, assistant secretary for the

USAF had approached George Mueller, NASA manager of manned space flight, to state the USAF's need for a high cross range and large payload capacity for the shuttle. It was a blunt threat not to support the shuttle unless the USAF got what it wanted. This pushed NASA over the fence and they finally, abandoned the Faget straight wing design and project.

The STS committee anticipated that the shuttle, now known as the STS (Shuttle Transportation System) would give the U.S. an "economical capability for delivery of payloads of men, equipment, supplies and other spacecraft like satellites to and from "space." The STS committee reviewed the NASA and USAF needs. The recommendation of the committee concluded such matters as "development and operational aspects, technology status and needs consideration along with interagency relationships."

FDL's concerns were sent to Myers who, by his own inclinations and his own familiarity with criticism of the design from within NASA was already predisposed to favor a delta wing approach. LeRoy Day recalled from a 1983 interview; "...For a long time we were still continuing to argue about this low cross range, or straight wing versus the delta wing configuration and Dale was the one, I think, that really held the line and said, "No, we're not going to go for this straight wing business. We're going to go for delta wing vehicle." Day went on to say; "And by that time, there was lots and lots of evidence from experimentation done within and outside NASA and the straight wing configuration had so many limitations that we really ought not to embark on that. We ought to build a hypersonic that was, in fact, a hypersonic vehicle and that meant that it had to be a delta wing." While Faget's concept had its devotees, it was seen in later studies made after 1971, when NASA went for the delta configuration was the prime shuttle candidate.

The 1971 version would have been shorter than a DC-9, two engines in a tailcone with a de-orbit engine and having one engine located in the nose behind the swinging nose cap with a conventional or T tail. NASA called it the "Blue Goose". It appeared in 1970 and disappeared just as fast.

Due to some very complicated heating issues with the Blue Goose, the configuration was wrong from the get go. At first, NASA was looking at variable geometry canards at the nose. It was dropped in favor of a sliding mechanism, so that wing could move twelve feet, with hopes that the wing could move in and out in orbit so no horsepower was needed. This sliding wing root would involve a very small structural penalty. After all the testing, it was finally dumped again in favor of the delta wing. Yet, more time went into bickering, hemming, and hawing about what configuration when the determination for the delta wing was the way to go. Shuttle development in 1970-71 was a haze of studies, studies and more studies, many of them conflicting and repetitive.

Studies started in February 1969. PHASE A resulted in PHASE B proposal in June of 1970. The next month NASA gave PHASE B follow on contracts to McDonnell Douglas and North American Rockwell. At the time of the Phase B studies award, the agency contractors were pretty much in agreement that the design of a large two stage, fully reusable spacecraft with a fly back pilot booster and orbiter, was the way to go.

That would include payload and fuel carried internally. As a hedge, NASA also awarded PHASE B contracts and two additional PHASE A studies to Grumman/Boeing and Lockheed for examination of a partially expendable system as alternatives to the PHASE B already in evaluation.

However, from 1970 to 1971 major changes were going on which greatly affected the ultimate shuttle design.

The Shuttle and flight test

There was yet another proposal that was working its way through NASA and that was the "Big G" proposal, which called for a Gemini spacecraft type vehicle to haul twelve astronauts to an orbiting space station. This faded in 1969. It went down because it was a personnel carrier and not suitable as a high volume payload/transport. NASA was throwing everything it could at the STS problem. PHASE B now went into the PRIME status and soon followed by B Double Prime Phase before the award for PHASE C finally went to North American Rockwell in 1972.

All this time NASA continued in house shuttle studies using engineering design support from Marshall Space flight center in Alabama and JSC (Johnson Space Center) in Houston, Texas along with Langley in Hampton, Virginia. NASA now had established the "LEAD CENTER" concept. What that meant was the "urination on the wall" problem started here. In short, the many centers that NASA had established within itself were now fighting for important turf. The choice seemed to be for Marshall to get the engine development, the ET (external tank) and SRB (solid rocket booster) development, since Johnson Space Center had already taken over the orbiter. Other NASA centers took over tasks they could handle. Dryden got the planning of the approach and landing development while Kennedy Space Center of course, got the service and launch of the STS.

The continuing PHASE A and B contractor studies were running together and were a duplication of effort. NASA's concept for that had to do with was it helping internally with its staff expertise and putting them into a better position to monitor contractor efforts because they were looking at this as an "internal learning curve."

In a 1984 statement by Aviation historians, John Guilmartin and John Mauer of NASA; "We really trained our people on what the shuttle was all about, by doing these in house base line designs, and as a result, we came to appreciate some of the problems a whole lot better, so that when we worked with a contractor. We had a better feel for what had to be done and what the contractor should be doing for us." Considering that this statement was made in 1984 and Challenger happened in 1986, it is obvious NASA did not understand its contractor base as well as it thought.

JSC engineers were calling the new shuttle operations the "DC-3 studies" in reference to the Douglas DC-3 workhorse of the aviation world. George Low, a veteran from NACA, became deputy administrator for NASA in the end of 1969. He felt NASA's A and B studies were moving toward the totally reusable system.

Low questioned it and it was not the best they could do given the uncertainty of the risk. In 1970, NASA concluded it could not do both the shuttle and the space station. While the shuttle was a cheaper venue and the Office of Management and Budget (OMB) in 1971 expressed the fact, it was not willing to support NASA at the agency's current level of 3.2 billion. This made the shuttle's position precarious at best. While NASA still wanted a fully reusable system, it had to swallow a partially expendable booster system. The shuttle did survive both restructuring of the national space program and a secondary blast from the OMB to support from the Department of Defense in the name of the USAF. This support from the DOD gave the USAF a big portion of the shuttle design. There were five main issues affecting the shuttle program:

- Determining capacity and dimension of the payload bay.
- Determining an optimum cross range.
- Choosing a TSTO (thrust augmented orbiter shuttle).
- Delete the plans to incorporate a landing engine.
- Select aluminum as the primary structural material.

The first two of the problems came from the USAF who was fighting for that big payload bay and the long cross range, so it could make polar orbits to launch its large satellites. These were critical issues and technical fall out from these requests, along with external factors that came flying out between the USAF and NASA. The 1960s and 70s showed NASA with two different space programs, one civilian and one military. With that, you had the many committees like the ACCB and the STS committee to add into it. From the beginning of Leroy Day's SSTG (Space Shuttle Task Group) had included both the USAF and the DOD representation resulting in a series of classified documents on the future DOD/NASA relations on the shuttle.

The USAF had no money to put into the shuttle development. Since the early 1970s, it found itself in the process of force restructure and that meant it was rebuilding many of its jet fighters and ordering new ones. The USAF wanted to operate the shuttle on military missions including the polar orbit launches from Vandenberg AFB. The USAF was willing to support the shuttle in Congressional hearings provided it got its digs in for the defense community. NASA felt the support was crucial to the STS if it were to withstand the Congressional attacks from both parties questioning the space program's need and rationale. However, on the one side of Congress, there were certain Congressmen who had different feelings about going into space, mainly, why are we going into space at all when we should be cleaning up the problems of hunger and poverty right here in the United States. The problem was not only a matter what program should be funded; it was a matter of federal versus state when it came to funding. NASA was not concerned about the USAF's lack of funding to support the shuttle program development. Having just NASA responsible for the shuttle funding really did simplify things like the budgetary approval process in Congress and eliminated the danger of having a program dependant on sources of funding with double the danger of budget cuts. With the 1970s, Congressional opposition to space in general and dislike for the shuttle in particular, it gradually failed allowing NASA to proceed with shuttle project without the stress of significant Congressional opposition.

The USAF support did balance the shuttle program in three ways.

- Payload capacity
- The bay size of the orbiter
- The cross range of the orbiter

The USAF had established the payload size of 15ft. x 60ft. diameter and the payload capacity of 65,000 lbs. It also allowed for cross range of 1500 nautical miles, which was later, reduced to 1100 nautical miles. The first two items were needed because the Department Of Defense planned operational systems. The shuttle needed to return to Vandenberg AFB after its polar orbit and needed the cross range to do it. In a major meeting held between NASA and the USAF, the payload bay and the cross range were a major factor.

The USAF added another significant input to NASA on the shuttle that of payload bay structural criteria. NASA wanted 9G's crash load while the USAF wanted more like 11G's. NASA revised the crash load criteria to a realistic level based on the shuttle's anticipated 3.3G's ascent load and this change was made to make it easier for the payload bay to develop. Overall, the squabble about payload bay and cross range held NASA to the fence with the USAF.

Like it or not, NASA had to accept these terms but not without another criteria argument, that being the delta wing which the USAF was against. The USAF wanted to have the Vandenberg AFB and launch site access. According to LeRoy Day: "…If you were making a polar type launch out of Vandenberg, and you had Max Faget's straight wing vehicle, there was no place you could go. You'd be in the water when you came back." The NASA design division kept up with the straight wing orbiter right through December 1971. However, when the decision was made on the mission requirements and the recognition of the problems with straight wings was beaten to the ground, the straight wing was finally out and the delta wing was confirmed. Also confirmed in 1971, was the ultimate configuration for the shuttle and the fact it would carry both of its propellants. Yet, the design of the orbiter was one task. The other was the external configuration and the whole vehicle with external propulsion or tanks.

The configuration called the 040 and dated, February 18,1972 now had the look and feel of the shuttle as we knew it.

As the PHASE B PRIME study went on, decisions on propellant and where to place it where it would be solid or liquid and whether the shuttle would use sequential or parallel boosters still had to be figured out.

Does it ever stop?

One of the last problems with the large payload bay shuttle came when the Office of Management and Budget (OMB) decided in the fall of 1971, that it wanted NASA to look at the smaller payload bay of 10 ft. x 30 ft. and evaluate it. A shorter payload bay made the balance of the shuttle more difficult. Finally, NASA succeeded in convincing Congress that the larger payload bay was the way to go and made if official on December 29, 1971, putting an end to the issue. By January of 1972, NASA had spent $91,749,000 on shuttle development. On January 3, 1972, NASA received the White House authorization to go ahead with the development of the shuttle and two days later, President Nixon publicly endorsed it.

A very BIG wish list

By 1976, NASA was looking at the drop test of the shuttle orbiter and flight test by 1978. The Department of Defense was looking for at least five hundred missions over the next twelve years.

NASA considered this a conservative estimate. There were still questions to be answered. It was not until 1972 that NASA finally figured out the solid and liquid propellant problems. As of July 25, 1972, a joint NASA /USAF source evaluation board selected North American Rockwell to develop the orbiter after a very long process. Since the shuttle was the "only game in town", with the end of Apollo, James Fletcher, NASA Administrator, summarized the board's findings.

The Source Evaluation Board leaned on expert advice using panels and total number of NASA members. The Board numbers evolved in its proposal review process which now came up to 416. The Board's proposal:

- Manufacturing test and flight test support
- Subsystem engineering
- System engineering and integration
- Maintainability and ground operations
- Key personnel and organizational executives
- Management approaches and techniques
- Procurements approach and techniques.

With this basis to move on North American Rockwell came in first, Grumman second, McDonnell Douglas third, and Lockheed fourth in the scoring process. George Low and James Fletcher stated: North American received the highest score in mission suitability and an overall rating in the good to very good range. The North American design provided the lightest dry weight of any of design submitted.

For guidance, navigation and control, North American used a triple redundant, single string approach, which the board considered very good as a simple design with minimum interference. It's good understanding of all electrical power subsystems reflected the very thorough studies that North American made following the Apollo 13 accident which had its origin in an electrical subsystem. The board considered North American choice of a male/female concept for docking to be less advantageous than the androgynous method proposed by other companies.

North American had the best of advantages over the others in the mission sustainability area and management. The plan showed efficient, centralized control by the program with a chief engineer and deputy. The Board felt North American had top project management and plenty of experience in manned space flight.

The final word and her name was Enterprise

NASA had another vehicle type up its sleeve; it was one of the many designs that floated around at the time. Vehicle 2A, also known as the "150K Orbiter" (K=1000lbs), brought about the largest change of the entire developing shuttle.

The vehicle's dry weight changed the design so much that the orbiter was resized. This included a wing twist to the camber and revision for improved subsonic performance. Improved low speed performance was due to the reduction, which changed the wing size of 269 ft. and re-balanced the orbiter for stability. North American Rockwell had the final configuration and NASA and Rockwell were ready to start the building of the first shuttle.

Orbiter 101 (OV101) named "Enterprise" began construction on June 4, 1972. While the Enterprise lacked many of the features of a fully complete orbiter, it was a full scale flying mockup.

OV-102 named Columbia didn't fly until 1981, three years after its planned initial orbit and missions due to unexpected developmental problems with the SSME (space shuttle main engines) and the TPS (thermal protection system). With Enterprise's completion, the shuttle program moved from the drawing board to the plant floor. Enterprise would be the first to fly, albeit in the atmosphere, which would validate approach and landing tests. 1977 saw the shuttle's approach and landing tests. By Fall 1974, the USAF and NASA executed a joint agreement to establish shuttle support and test facilities at Edwards AFB including the construction of the mate and demate facility which allowed for the new 747 Boeing transport jet to carry the shuttle on its back.

NASA's Johnson Space Center would have the overall mission control authority. The Air Force Flight Test Center (AFFTC) and Edwards would also provide technical support and analysis as needed. NASA would arrange for shipping the shuttle from the construction site at Air Force Plant 42 at Palmdale, California and travel forty miles overland to Edwards AFB which allowed for the construction of a special "shuttle road". OV101 rolled out of its Palmdale plant on September 17, 1976. The flight test program had three phases:

- Captive
- Captive-free
- Free flight.

The unmanned captive flight would show if the 747 Carrier aircraft and the shuttle could fly together comfortably. Boeing added a vertical fin to the 747's horizontal stabilizer. The shuttle launched on the 747 with a tailcone and the flying tests couldn't be more perfect. By 1977, the shuttle was on the way. Rockwell and NASA initially took off the tail cone and replaced it with something equal to what a shuttle would have on reentry from space, including the three main engines nozzles and smaller nozzles of the OMS system (Orbital Maneuvering System).

The flight on October 13, 1977 was again perfect. The October 29, 1977 flight at Edwards did show some issues with control at touchdown. The flight tests continued to correct it.

On March 10, 1977, Enterprise left the Dryden Center at Edwards AFB for the last time. Flown by Fitz Fulton of XB-70 fame and his crew, they headed for Ellington AFB in Houston, Texas where the Orbiter/SCA was presented at the air show and saw some 240,000 visitors. On March 13, 1978, Enterprise left for Huntsville, Alabama and NASA's Marshall Space Flight Center for ground vibration tests. There were three major issues with the shuttle between approach and landing in 1977 and its first orbital test flight which was scheduled later in April 1981.

Resolving the flight control system problems that the shuttle found on the last tailcone flight, the developing of the SSME (Space Shuttle Main Engines) and the TPS (Tile Protection System) system were still major hurdles to climb.

The first fix involved a software change to the digital flight control system, which was at the time considered, a big problem. It would take care of the approach and landing problems. The second problem of the SSME and the third of the TPS system were major headaches, indeed. The SSME by 1979, after much testing was looking better and had sustained 91,000 seconds of test stand time with the abort mission simulation runs of 665 and 823 seconds, which was compared to 520 seconds needed for orbital insertion.

Nevertheless, in 1980, the old problems came back with a vengeance. In July 1980, three SSMEs failed during two weeks of trials at the NSTL (National Space Technology Labs in Mississippi). In November 1980, a weak brazing section on the nozzle of the SSME gave way during a 581 second engine burn which punched a hole several inches in diameter through the nozzle. This resulted in a destroyed component but not an entire loss. There was extensive work done on the SSMEs and work resumed December 4, 1980 with successful firing of the SSME at 100% thrust for 591 seconds.

It's not over yet

Early 1981 found NASA thinking it had solved the worst of the problems with the SSMEs. The engines that everyone took for granted, the SRBs (solid rocket boosters) was the one who had the serious and destructive flaw. It was the SRB that fell through the cracks at quality control, and that along with the very poor management system led to the death of seven astronauts and the shuttle Challenger.

The third problem of the TPS (Thermal Protection System) was still a big headache. Back in the 1960s, proponents of the shuttle had argued the question of active cooling systems, which meant either using fluids for heat sink or passive cooling techniques. The passive cooling techniques or system weighed less and was less complex. However, refurbished ablators (like those on the X-15-2 space plane) showed some promise but the problems of refurbishment quick enough to satisfy an ambitious program, wasn't overcome. The problem and the TPS engineers soon found themselves looking at "hot structures" made from some sort of exotic material, which seemed to be the only answer. Using a product made of some sort of ceramic tile or coating attached to the orbiter's aluminum frame is what NASA finally went with. In an early PHASE B study, NASA selected ceramic tiles over metallic shingles and confirmed it in February 1973. The decision to use a non-metallic heat shield on the orbiter was made. This decision caused a competitive evolution of *silica* and *mullite* (aluminum silicate). Silica won the day due to high thermal performance. NASA engineers like working with aluminum since, for them, it was a know quantity unlike titanium and all of its alloys. Lockheed in its *Star Clipper* design was using aluminum, which was odd since Lockheed was the forerunner in working with titanium, as seen in its very successful A-12 Blackbird and the SR-71.

It was a very strong statement having a company so versatile in titanium and mixing alloys, to design its orbiter in aluminum. Lockheed also created the ceramic reusable surface insulation (RSI coatings LI-900 and LI2200) with a difference in density. The LI-900 had a density of nine lbs. per cubic foot; the LI-2200 had a density of 22 lbs per cubic foot. These soon became the CLASSII (LI-900) and CLASS III (LI-2200) insulators for the orbiter. The ceramic media was made of silica fibers that was held together with other silica fibers and glazed by a reaction cured glass consisting of silica, boron oxide and silicon tetra boride. This mix was not waterproof. A silica polymer was coated on the undersurface (not glazed) side of the vehicle. The coating was very brittle and Lockheed could not coat the entire vehicle with it. Instead, the coating was made into smaller tiles and these tiles would have small gaps between them (less than 1/1000th of an inch) to allow relative motion and deformation on the metal structure due to thermal expansion and contraction.

There was another problem concerning the local expansion of the aluminum skin directly under each individual tile, since a tile could crack under the load. A decision was made to isolate the aluminum skin from the tiles by bonding the tile to a felt pad and coating the felt pad to the skin using a silicone adhesive cured at room temperature. It seemed perfect for the shuttle problem. It created a tile structure that could radiate heat and insulate the orbiter from heat of reentry. However, the tiles were always fragile and critical at best. Due to yet another problem, resulting from voids in the foam covering the external tank for insulation, at ignition this foam would break off the tank and hit the orbiter such as the case was with the loss of Columbia and seven astronauts. This problem of the foam was something that was in the first flight and the last flight of the shuttle. It was never truly corrected.

The shuttle used four levels of reusable surface insulation, two of which were the tiles. The high heat areas of the shuttle, such as the nose, cap and leading edges, were protected by RCC or reinforced carbon-carbon.

The temperature would range between 2300 degrees and 2700 degrees, well above the 350 degrees maximum temperature, which the shuttle's aluminum skin could tolerate. On areas where the temperature could range from 2300-2700 degrees F., the area would receive high temperature reusable surface insulation (HSRI) meaning that most of the orbiter would get these tiles (LI-900) or the smaller (LI-2200) which gave the shuttle its black belly. Areas with temperatures up to 700-1800 degrees F would receive the white tiles (LI-900) called the low temperature reusable surface insulation (LRSI). This would blanket the area where the temperature would not go above 750 degrees F. during ascent or reentry.

NASA was getting quite nervous regarding the direction of the shuttle program. It had initially thought it could solve the TPS protection problem of the lee side of the shuttle. However, in March 1975, a NASA panel and FDL engineers showed NASA otherwise. In April 1978, an AFFTC/FDL team worked at the potential orbiter heating problems and in January 1979, the panel managed to isolate the problem. The OMS (orbital maneuvering system) pod area showed to be a thermal heating problem, which did show up in the orbital test flights. To support the FDL findings, the USAF ran tests using wind tunnels at the AEDC from May through November 1980, at the Naval Surface Weapons Center Tunnel in May of 1980.

Tile Problems

At this time, a critical problem showed in the TPS system (thermal protection system), it was the tiles themselves. The shuttle was getting closer to operational flight test studies and both NASA and North American Rockwell had concerns regarding the loads the shuttle would have to face in flight. It was clear in 1979 that certain areas of the shuttle TPS system could not withstand the load of a normal mission. NASA was worried the tiles would come loose because of the loads caused by flight and they could drop off from the shuttle during reentry leading to a "zipper effect" where the tiles would shed off the aluminum structure and into catastrophic failure. The agency went on a manhunt to find the answer. That meant NASA put the call out to academics, government agencies, industry and outside panels of experts. Le Roy Day, leader of the Shuttle Task Group said: "...there is a case (the tile crisis) where not enough engineering work was done early enough in the program to understand in detail—the mechanical properties of this strange material that we're using was neither fish nor fowl..." NASA's push to find the answer to the tile problem worked but caused yet another delay. The solution involved the strengthening of the bonding between tiles and the felt strain pads that were used underneath. The analysis showed that with each tile, the SIP (strain isolation pad) under the tile and the adhesive between the SIP and the aluminum skin all had the recommended strength individually, however when put together as a system, it had a reduced tensile strength of overall 50%.

In October 1979, NASA had gone for a 'densification" process to fill the voids between the fiber as the location of the bonding surface with a special slurry mixture call LUDOX (made by Dupont Corp) made of colloidal silica and a mix of water and silica. With both air and oven drying, along with waterproofing, (exposure to fumes of methyltrimeoxysilane), and a Dow Corning product called 7206070 and acetic acid, the newly "densified" tile was applied. However, insulation was tedious and extremely time consuming. By March 1979, the shuttle Columbia flew across the country on the back of the 747 and arrived at Kennedy Space Center in Florida. The shuttle then spent twenty months, with a three-shift work force that went six days a week to apply the 30,759 tiles to the shuttle. Each worker would place at least 13 tiles during their shift. By mid 1979, along with workforce strike the year before that stopped production cold and adding the new "densification" process, the tile problem was finally under control by early 1980.

Three years later than NASA planned in March 1981, the target for launch went into place. James Fletcher, then NASA administrator had moved on and was replaced by Robert Frosch. Frosch made it clear there would be no more slips for launch date.

Frosch, in his wisdom, decided to "densify" only half of the tiles recommended, taking care of only the critical ones after the first flight. It did work but it was a case of "Russian Roulette" as far as the shuttle was concerned. Because of Frosch's hurry, the shuttle Columbia went into the VAB (Vehicle Assembly Building) at KSC on schedule, November 24, 1980.

The NASA Centers and the Fight for Political Position

One of the main issues that plagued NASA since its inception has been the politics that has wormed its way throughout the entire NASA system of centers. It's not a brand new headline that the various NASA centers have been vying for not only a piece of the project, but when they do get a piece of the project, whatever it may be, it becomes hoarded treasure, sacred to that center.

As far back as the creation of the Kennedy Space Center and Cape Canaveral, the fight for NASA's autonomy from the USAF has created many an internal war. That "war" crept into the other centers namely Marshall Space Flight Center in Alabama, the home of Werner Von Braun and his missile system, right to Manned Space Flight Center now known as the Johnson Space Flight Center. Even the creation of each of NASA's home bases, has been a political pork barrel item. It is one of the main reasons that NASA broke down in the Challenger and Columbia disasters. To get to the meat of the matter, we need to look back at the creation of the three main NASA centers to see just what happened.

The NASA Centers or How to build a Launch Site

The inception of the Kennedy Space Center started was back when the USAF was using the Cape Canaveral Air Station as the launching point for many of its own missile tests. In 1959, a plot of land outside of Washington D. C. owned by the Department of Agriculture, became the Goddard Space Flight Center.

The original role of the Goddard Center was to bring together NASA and all it abilities from the existing centers and add personnel for operating space missions. However, human space flight and the moon program called Apollo spurred the creation of yet another new NASA Center a few years later. NASA and the Manned Space Flight Center about 25 miles outside of Houston, Texas (later to become the Johnson Space Flight Center) would take on the responsibility for mission control and the training of the astronauts along with other flight related activities for deep space manned missions and of course the golden dream of APOLLO. Goddard's role was scaled back to scientific investigations through near earth flight without human crews, The JPL or Jet Propulsion Laboratory in Pasadena, California would be responsible to monitoring the robotic flights of the probes sent out to the moon and beyond. Now we now find ourselves looking at one of the most decisive parts of the NASA Centers. In 1957, Kurt DeBus was the director of the Cape Canaveral Center and a launch director for the NASA. He worked with the Redstone Rocket team and served as a launch expert. Even back then, it was obvious that there should be a division of "church and state" and DeBus felt that the Marshall Space Flight Center should use its resources and not pay the USAF for the privilege of launch services.

By 1959, when T. Keith Glennan, the first of the NASA administrators established an office at Cape Canaveral to use as a go between for NASA and the military commander, General Leighton Davis of the USAF; it was one way to keep an eye on NASA's projects that were using the launch facility. The office was under the control of the Director of Space Flight Operations, AMROO that would over see the project offices that were assigned to specific project and liaison with the USAF and their range officers.

This process did not last for long. While AMROO[5] had to "recognize" the responsibility and authority of heads of divisions and offices, it was more a coordinating service than an operational one. This showed

---

[5] AMROO—Atlantic Missile Range Operations Office

the "early on territory infighting" that began within each division of NASA.

DeBus worked on the early space programs, Mercury, Gemini, and the other launches that went off at Canaveral. On August 24, 1961, NASA administrator James Webb and Deputy Secretary of Defense Roswell Gilpatric signed an agreement on the management and funding of the Apollo lunar launch program. In that same year, DeBus and General Leighton Davis were asked to prepare a comprehensive study of launch facilities, starting with the most basic question of location. The new agreement with the military and all the investment and experience at Cape Canaveral, it still wasn't written in stone that this would be the place for the NASA launch operations.

The Cape's relative isolation did have advantages especially in the early years when launches blew up on the pad. With the idea of protecting the public and the flat, remote landscape, the Cape offered a clear line of sight between the operations center and the launch pads, which needed to be separated because of the explosion possibility. The few miles of separation between the blockhouse and the launch pads would not do with engines capable of generating millions of pounds of thrust.

DeBus and Davis looked at other options as well.

In July 1961, there were many locations under consideration:

- Mahayana island—Bahamas
- White Sands Missile Range—New Mexico
- Christmas Island—Pacific Ocean
- South Point—Hawaii
- Location Brownsville Texas, possibility of an offshore platform in the Atlantic Ocean.

However, Canaveral was the first on the list. It had everything needed including people and buildings. The local populace were used to the noise and "Rocket Racket." It would be easy to build Canaveral up for more launches and there was room to grow on nearby Merritt Island.

The Atlantic Missile Range, some 1600 miles downrange from the tracking stations in the Bahamas, Brazil and the Dominican Republic, offered a vast field for flying missiles. Still, the Cape had enough disadvantages to keep the final choice uncertain. It would be expensive. Labor was a question with few skilled workers in Florida, which was non-industrial and short on basic materials like copper wire, cables and machinery like transformers. Weather was a big issue as hurricanes and storms were always a threat. No one really knew what a hurricane would do to the launch facilities and was later to find out just what cold weather and ice would do.

Bureaucratic politics also had a role in the choice of Canaveral, along with the USAF, who already expressed interest in running launch operations for NASA and itself. To Kurt DeBus, it presented a danger. NASA's launch center might be "swallowed" by existing organizations. Possibly leaving the Cape to the USAF and getting away from the military facility was better. At least NASA's independence and autonomy would now be a factor in helping it design its own launch facilities.

In the end, availability of tracking and range facilities and existing launch complexes pushed the decision in favor of the Cape after all. NASA announced August 24, 1961 that it would purchase some 88,000 acres of adjacent land to Cape Canaveral Air Station on Merritt Island at a cost of $72 million. Merritt Island was located between the Indian and Banana Rivers and the site would provide space for a permanent launch operations center. With the location settled, NASA now returned to the organizational questions at hand. Marshall Space Flight Center and the Launch Operations Directorate grew significantly over the years. In 1953,

DeBus had 19 people to oversee. In Huntsville by 1960, he had over 500 people.

There were three options drawn up by DeBus in June of 1960 that were still on the table. It was clear that simply retaining a Marshall directorate for launches at the Cape was not going to work. The decentralized approach was still attractive, allowing for a small central staff that could coordinate separate launch divisions controlled by Marshall, JPL, Goddard and the Manned Space Flight Center in Houston. Under this proposal, DeBus would wear two hats: one as central coordinator of NASA launches and two, as launch director for Marshall. There were many at Marshall who liked this option because it allowed them to retain their own launch schedule and organization separate from the Cape. The advantage was that it involved the least change from a personnel and organizational stand point and conserved the experience already gained. It kept control of the Cape on Marshall's side.

As you can begin to see, this is a basic root problem when it came to the organization of the centers. It starts right here at the inception of the Cape and how to divide the goods at hand. DeBus saw very early in the game just what could happen with politics and infighting between the centers. By October, the debate tapered down to two options. Whichever center won the launch authority; it would be separate from any other center. Historic ties between the Cape and Marshall notwithstanding.

The decentralized pattern option remained on the table. This proposal was strong: allowing for an in-house ability and the workbench proficiency DeBus wanted, while allowing personnel to get experience, to move launch operations up the learning curve quickly. This approach could be criticized for necessitating duplication of equipment and making coordination hard. There would have to be some sort of mechanism to feed the accumulation of knowledge from the various launches across the NASA centers.

DeBus and Werner Von Braun, (who was the head of Marshall) now fought for a single central launch division even if it couldn't stay within the Marshall Space Flight Center structure. They had to focus on NASA's number one priority, the new lunar mission called Apollo. The program was going to take up time, money and people. A fully independent center had to be independent enough to serve its major user—the Apollo mission— without unreasonably impeding other programs or missions.

There was one way to do this and that was to maintain the existing structure for the ongoing Mercury/Atlas program. This would compromise increased support for a central independent launch authority, just the thing that DeBus and Von Braun wanted.

The concept had central planning and execution and kept NASA in control of NASA decisions by meeting requirements and operational characteristics of a "NASA facility". This was the plan NASA approved on March 7, 1962. DeBus became head of NASA Launch Operations. The Cape would now be separate from Marshall's Launch Operations Center and would serve all vehicles at the Cape with special arrangements for some projects. DeBus reported to Brainard Holmes, the Director of Manned Spaceflight and NASA's Washington, D.C. Headquarters.

However, DeBus kept his "second hat" and was in control of Marshall's main project, the new, huge, Saturn booster that Von Braun was cooking up for the Apollo mission. The close connection between the vehicle work and launches required more organizational lines at least until the LOC (Launch Operations Control) at Canaveral gained enough experience and built its own connections back to the vehicle center. This separation agreement which severed the LOC of Cape Canaveral from the Marshall Space Flight Center was signed June 8,1962.

On July 1, 1975, Marshall employees started to work for DeBus on paper. NASA now had a launch operation center (LOC) but this organization didn't have facilities of its own at the Cape and worked at the Huntsville, Alabama location. This might all seem terribly confusing and in truth, it most likely is. However, what this does show is how the initial power grab between NASA centers, even the early, first centers like Marshall and Cape Canaveral which had to wrench themselves from the grip of the USAF when it came to launching rockets. What we need to remember is that the USAF was already in the missile launch business and they wanted to control the civilian part of the launch program, too. However, Kurt DeBus had seen into that and wanted autonomy for NASA and the right to launch missiles and programs under *their* requirements, without having to ask the USAF for permission to use the launch pad,

Two hats in the ring

While DeBus split his time between Alabama and Florida, working in offices borrowed from the USAF at the Cape, some of the other people launching vehicles were still part of the NASA/Mercury program that belonged to Manned Spaceflight Center which was home in Houston, Texas, better known as MSC (later the Johnson Spaceflight Center).

At this point, DeBus had no formal place in that organization. Mercury and Project Vanguard were out of DeBus's jurisdiction. Again, the division of responsibility between NASA and the military had yet to be finalized. New launch pads and equipment had to be built. It wasn't clear what DeBus's operation should be called. NASA headquarters in Washington D.C. suggested "National Space Pad", "National Space Operations Base", but LOC (Launch Operations Center) would have to do for now. Many of DeBus's people came from different backgrounds and not all adjusted to working with him. DeBus came from one of the most distinctive cultures within the already diverse NASA culture. It would take a master of a bureaucratic politician to bring all the groups involved together to begin building a spaceport organization.

1. The LOC (Launch Operations Complex) was responsible for planning and supervision of integration

2. Test checks and launch of vehicles

3. Launch concepts which would be controlled in "master plans"

4. Master Plan would be equal to a "flexible spaceport".

Joint panels would act as "law making bodies" within the LOC serving as the executors to enforce the laws. This would build consensus on matters that cut across different factions and didn't fall along lines of clear authority. There were relations with the USAF to work out. It was natural for the USAF to view the NASA facility as an extension of its own missile test site, or at least that's what they would like to have happened. The Atlantic Missile Test Range (AMROO) always had tenants on its property to launch missiles. This was the approach envisioned when NASA administrator James Webb, and Ross Gilpatric (Secretary of Defense) signed the 1962 agreement. Recognizing that new launch facilities would be needed for Apollo, the agreement stipulated "it would be in the national interest for NASA and the DoD (i.e. USAF) to cooperate". The new launch facilities "peculiar to space program" would be important in fixing the rate at which the program would proceed. The launch site was jointly managed through NASA and would spend its own money for the new facility. It would also give the USAF some key management responsibilities, including range, flight and safety management along with master planning of the new facilities.

NASA's new relationship

As we dig further into how relationships developed in the early stages of NASA centers, we need to take a look at how these relationships transformed the centers and what caused the centers to dig in to protect their own "piece of turf" instead of working as a team. As NASA and the Apollo project grew, the relationship between the USAF and NASA was out of kilter completely. The new plans for the LOC (Launch Operations Center) included acquiring and building new structures in the USAF's backyard, which would have been on the west side of the Banana River.

Military spacecraft would use the same range proposed facilities on Merritt Island. This would be under NASA control and designed for NASA missions. Problems would arise when it came down to sharing facilities and space. DeBus wanted to be sure that LOC had its own identity. This required a true joint plan with neither LOC nor AMROO making unilateral decisions. With NASA working on its own decisions for design of the new Saturn V rocket and unmanned craft as well, it would fall on "existing DoD regulations". This meant that decisions on range and launches might not be sufficient for NASA.

Land Grabs

It was assumed that a master planning board with both USAF and NASA might resolve the problems. However, AMROO had another idea. General Leighton Davis proposed that the USAF and not NASA should own the land and facilities that were being contemplated for Merritt Island. This was a major change; one that would have ended NASA's LOC plans for good and all, before it ever got started. The USAF's plan was obvious. It ran the range now and was best placed to handle the new facilities. Smooth integration would come with their supreme authority.

On March 27, 1962, General Leighton Davis told Kurt DeBus he wanted to follow the "Webb- Gilpatric" agreement that would make the

USAF single manager of the entire Cape area. The USAF, Davis argued, should hold title to the new Merritt Island land. General Davis was ready to go to Congress with this. However, DeBus objected strongly. Only if NASA held its own land could joint participation with the DoD in planning and use of the AMROO would work. Otherwise, all improvements, including NASA's mission facilities, would become USAF property. All involved knew that the mission to the moon was not going to be a one shot deal. NASA was not going to be a temporary resident on a military facility. However, creation of NASA was the start of a new civilian era in space and it needed the freedom to advance at its own pace and build its own launch facilities. It was crucial to the civilian and scientific nature of the missions that this be "clear in the public eye" which would be harder with the military in charge. NASA would still be a small customer in a large military operation.

Using the original NASA motivation, DeBus stood tough. NASA was a civilian agency and had to have its own property at the Cape and freedom to build its own facilities. Both Secretary of Defense McNamara and James Webb, NASA administrator agreed Merritt Island Launch Facility was NASA's and LOC became the single manager of the new facility. NASA would make policy and have full power to determine launch procedure. There would be arguments as to whether each side was living up to their part of the agreement but the Webb-McNamara agreement of 1963 stuck in place.

This all meant that LOC (Launch Operations Control) would now have full power to determine how launch operations would happen. LOC was now NASA's single launch agency at the Merritt Island facility.

All the encompassing functions like, public relations, community relations, visitors services, purchasing contractors, law and security would be NASA's. NASA took this very seriously. At least now, NASA was able to make all its own local arrangements without approval by NASA HQ.

Kurt Debus still fought for a separate organization to handle launch operations. DeBus' experience afforded him the insight that concluded that not many of the designers would ever be finished with making adjustments to vehicles right up to countdown.

However, there was never anything like this separation before. To keep design and launch separation new provisions needed to be set up. These new development centers would be "responsive to the LOC's requirements in discharging its overall integration test checkout and launch responsibilities." The Marshall, Goddard and the Manned Space-craft Center in Houston (soon to be Johnson Space Center) would give "functional requirements" for LOC, which would allow for coordination and meet the needs (changes) of the various centers and contractors working on the vehicle. LOC's job was to keep the whole thing moving.

It all looked great on paper but DeBus was only one part of a huge complex with other NASA personnel reporting to their home centers. There were many programs going on from Langley, Goddard, etc. However, it would be the moon project that would finally shape all launch operations. With the creation of the lunar program, the cohesive-ness that became the NASA of the 1960s was born. As Mercury and Gemini, programs flew successfully and closed out, the push to the moon was on. The plan to get in all the centers and make them into one defining system was on. However, things were not quite that simple. Here we start to see the drive of the various NASA field centers to try to take as much work and prestige from the Apollo program that they could.

The Moon and NASA's Roles

President Kennedy was assassinated November 22, 1963. With the death of this beloved president, the affirmation to reach the moon at the end of the decade was in full force.

Congress was doling out the funding. NASA's budget was larger than ever. Cape Kennedy, as Canaveral was now called, was in full running mode. All of NASA was working on the lunar problems. Tests were being made and designs were in constant flux. There were labor problems with all the new contractors now working at the Cape.

The U.S. Air Force was already allowing union workers to work right on the missile itself. This was highly irregular. NASA wasn't too pleased with that decision. The Launch Operations Control was used to the Marshall Center and Von Braun way of doing things, which recalled its "in house" tradition. As we all know, the USAF and NASA never really saw eye to eye on many things. USAF was not pleased with the fact that now there was a Civilian agency running the space game instead of them. Kurt DeBus's statement: "We have always followed the practice to contract… with private enterprise for the construction of the basic launch facility. It was important to retain for our own employees that part of ground support, which is still undergoing development, and is a critical part of the missile system. LOC to DeBus was still a research facility and DeBus felt union workers could not respond to the many developmental changes, which were taking place during installation and checkout of a facility. Let's face it; the Launch Complex where rockets stood wasn't just a piece of concrete in the middle of an island.

One of the largest issues had to do with DeBus's concept that civil servants and engineers could do the work of union workers. Of course, that didn't go over too well with the union workers, who started to complain about the "food being taken out of their mouths". There were stands taken by the union workers when NASA workers would install a cable and union workers would cut and resplice it. This blessing of equipment by union workers would have them doing nothing to the equipment but maintaining the area as their own. Much criticism from the Congress was out there, because they were getting wind of what was going on at the Cape as to workers "riding the gravy train" by slowing down on jobs and requiring overtime and doing a sloppy job to boot. It wasn't just union rituals, there were others who saw how the USAF was handling things and argued there was no reason why union members couldn't install equipment. DeBus kept bringing up the concept of "Space Technology" was different from anything else. Some quickly evolving systems changed literally overnight and required reworking of a system that was completed the day before.

There was a shortage of supervisors and tools that meant some workers were just standing around without supervision. The question of jurisdiction was another big issue in the debate of what belonged to Launch Operations and the space program itself, or did it actually have a goal at all.

With the amount of unsubstantiated territory, the relations between the contractors and NASA deteriorated. By 1964, job actions and strikes were becoming a way of life on the Cape. Probably the worst was a strike against the Florida East Coast Railway. While not against NASA, all kinds of pickets showed up at the gate and local teamsters refused to cross the picket lines. This meant that the Saturn launch vehicle schedule was thrown under the bus. The FECR (Florida East Coast Railroad) had earned its title as a very antiunion company. James Webb, who was the NASA Administrator, realized the danger to the NASA schedule. Webb tried hard to work with the railroad company executives to settle the "strike".

NASA's problem in all of this was the fact that the railroad had a line that ran on NASA land. However, Webb couldn't get them to budge. It took a judge's ruling to get work going again.

While this didn't end labor problems at the Cape, this action alone cost a total 1700 lost days of work. The strike of International Association of Mechanics against Boeing Aircraft (one of NASA's biggest contractors) cost still more days of lost work. The Cape eventually turned into the Kennedy Space Center in honor of the slain president, John F. Kennedy. NASA and the KSC were about to start its monumental roll into the Apollo program.

Much of this information about the early days of NASA was raised to show just how things formed early on and how NASA evolved. NASA's problems didn't begin with the Challenger disaster; it goes way back to the beginning of NASA. We are setting the stage for the object of our question, how and why the disasters happen.

Apollo AS-204/Apollo 1 and disaster

As the Apollo program continued in the midst of strikes and union problems, NASA was also to face its first major, manned space program tragedy. That is the Apollo AS-204 also known as Apollo 1 fire, which killed astronauts, Ed White, Gus Grissom and Roger Chaffee.

NASA, at this point, had started to work out its problems building the Apollo spacecraft and the Saturn V launch vehicle that would take them to the moon. All this came to be but not without enormous pain and loss.

By 1966, NASA Administrators were concerned that problems with all the vehicles would hamper the moon shot. The "coordination" between KSC and the new Houston Spacecraft Center now known as the Johnson Space Center were in a difficult situation.

It was obvious that trying to keep communication between both centers open to the Apollo spacecraft was a tough job. It was also obvious that trying to keep a running count of the lunar spacecraft parts was a huge project in itself, since mostly all of it was hand made. Moreover, of course, with the amount of contractors and subcontractors involved, innovation could be a nightmare for KSC, who was responsible for verification and testing of all the vehicles.

AS-204, which was Apollo 1, was already behind schedule because of issues with the Apollo craft that was used in tests. AS-204 didn't get to the pad until January 21, 1967, almost 6 months late.

Six days after delivery, there was a "Plugs out Test" on Pad 34. Here is where the disaster struck. During the test, the AS-204 Apollo Command Module, burst into flame and killed the three astronauts inside.

It was the first disaster of this magnitude to hit NASA right in the heart. Many felt that this would end the space program. There were myriads of investigations, investigators and recriminations, right up to accusations.

The loss of Apollo 1/ AS-204 was a spike at NASA's core. However, as we all know, it would not be the first spike.

What caused the Apollo 1 fire

Right after the accident, NASA called the AS-204 Review Board. Floyd Thompson, director of Langley Research for NASA was the chairperson. It took months of investigation, but the board found out a number of situations contributed to the accident.

The first was the pressurized, pure oxygen atmosphere in the capsule, wiring that was vulnerable along with no plan for crew egress in an emergency. These issues would come back to haunt NASA in later years.

One of the most important conclusions drawn was, no single person policy or mistake was held responsible. It was also the series of decisions that brought the new "risks into a non-hazardous condition". Pure oxygen in itself may not have been the heart of the problem, but the pressurization of the oxygen made it more dangerous.

Yet, a fire in a pure oxygen atmosphere never showed up in any NASA study. NASA associate administrator George Mueller said; "The fire had been considered remote, because of standards of design, manufacture tests and operations over the last several years." Yet, how could all these things accumulate without anyone noticing it? That means going back to the development of the Apollo craft and its problems with schedule, something that has always plagued NASA. Remember that those who do not learn from history are doomed to repeat it, as we will see with Challenger and Columbia.

Many schedule adjustments ended up overlapping the two Apollo - Saturn vehicles. AS-207 and AS-203 were unmanned flights. Because of the overlap, more people were hired and needed training to handle all the extra work so that AS-204, the first manned flight, would happen on time.

The contractors were already on the carpet for being late. By 1965, Apollo program directors, General Phillips had doubted North American Aviation's ability to deliver the goods. Phillips, after a long tough investigation, found that NASA's quality control was sub-standard. The quality inspectors were writing too many discrepancy reports, after the vehicle was delivered to the Cape. NASA's Quality Control inspectors found even more problems with the craft.

As the command module finally arrived at Kennedy Space Center in August 1966, AS-204 was unfit to fly. KSC technicians said the module had been shipped in a" peach basket". The program manager for Apollo, General John Shinkle, took position of the spacecraft with approximately 164 discrepancies and incomplete engineering orders. It only took one month for that number to grow to 377 change orders. Shinkle said 70% of the change orders should have been found at the contractor North American Aviation. While this was a horrendous state of affairs, it wasn't uncharacteristic. It was policy for contractors to ship vehicles unfinished to meet schedule and the KSC technicians would finish the job. Soon, this habit became part of the launch process. This turned into a game of catch up instead of a smooth work transition from contractor to NASA.

While AS-204 was getting cleaned up, the astronauts, White, Chaffee and Grissom were complaining that their "trainer" Apollo craft was behind the new update modifications that were coming in daily. Grissom actually hung a lemon by the trainer that expressed his view of the entire situation. Even so, the astronauts had faith it was all going to work out. However, others were not so positive. An inspector at North American Aviation by the name of Thomas Baron felt and warned of a very deep-seated problem with the quality control. His nickname became "DR" for Discrepancy report, since he wrote so many. Baron was frustrated by the fact his warnings were being ignored by the NAA officials. Finally, he just let some of the issues get to the press which eventually got him fired from his job at NAA.

Many of the Apollo issues were solved before the fire while others were not. For example, the environmental control system was replaced but it still didn't work. A communication system was so unclear it barely functioned. Yet, NASA continued to proceed with the next step, which meant the Apollo crew would check out the spacecraft in a "plugs out" test. After the three astronauts entered the capsule, it didn't take long for things to go wrong. The test was a lost effort from the start. Grissom complained when he hooked up an oxygen line about a foul smell. Alarms were going off periodically in the spacecraft. Communications was a flop as Grissom's conversations were broken up with the control room. There was a fifty-one minute hold while a communication failure was worked as they say, which meant getting it fixed.

It only took seconds after the resuming of the countdown that Roger Chaffee said he smelled smoke. Ed White quickly shouted behind him **"FIRE in the cockpit!"** The capsule soon ruptured and fire leaked out to the service structure. Workers on the pad jumped back, many ran for fire extinguishers, but it was too late as workers tried to wrestle the hatch cover open, only to find they didn't have the right tool.

Due to the changes in the Apollo spacecraft, Velcro, a highly common product, was extremely flammable in a pure oxygen atmosphere. That is what was pumped into the capsule. In seconds, the cabin was a frantic scene. The astronauts didn't stand a chance in the volatile atmosphere. The fire caused the pressure to spike, putting thousands of pounds of pressure against the inward opening hatch.

Minutes later in what seemed like hours, workers got the hatch open and what they found defies description. Two bodies were fused by their pressure suits on the cabin floor with the third still strapped to his seat with melted plastic dripping in the blazing heat. The flames didn't kill the astronauts; it was the smoke that was toxic, and it finished them in less than two minutes. The control room heard the last screams from the astronauts, completely helpless to aid them.

No one at NASA or anywhere else was ready for this. While some workers heard the astronauts "kid" about the way they could die in space, no one foresaw them dying this way. Many people reacted by crying outright, while other just stood in silent horror.

All of this caught NASA and the people of the United States right in the gut. This was the first death of astronauts. Astronauts had sixteen successful flights in space. It all seemed incredible that they could die on the ground in this fashion. Things like this didn't happen to the "CAN DO" NASA. Unfortunately, it did and the effect was harrowing. NASA was no longer the omnipotent giant of the space age. The press fueled the fire with sensational articles in tabloids and biased reporting. There were many accusations against NASA and the tabloids had a field day. Of course, the one source of criticism against the disaster was Thomas "DR." Baron, the ex-employee of North American Aviation. He had written a report on the poor workings at the Cape and at NAA. Why and how NASA fought back against these accusations was a useless effort. The Public already had the feeling that something was not right at NASA. Congress was also having its doubts. The Congressional investigation of the fire found much in the line of problems. Senator Walter Mondale used the "Phillips report" of 1965-66, which was very critical of NAA's work. Baron did testify in front of Congress. The strange and sad part of this is that both Baron and his family died in a car crash only a week after his testimony. Chaos and conspiracy theories ran wild.

In addition, NASA blamed NAA who blamed NASA who blamed designers and on it went. However, NASA did manage to fix the problems with the Apollo spacecraft and the return to flight was imminent.

Overall, Apollo was an unqualified success. However, the roadwork was laid for the different NASA centers that wanted their own piece of space pie. By the time the shuttle program came into play, the division between the centers was apparent and it would cause trouble in the very near future. By 1974 Kurt DeBus, director of KSC, made it known he was going to retire at an early date. DeBus in his decision said, "Our type of work, it's misery." For thirty years, the pressures of demanding programs have exacted their toll. Debus was proud of how he had developed Cape Canaveral and Merritt Island on the Kennedy Space Center. He decided to leave when; "We were well along in design and construction of facilities and supporting systems to accommodate the Space Shuttle which will open up a wider frontier in space and activities in the near future."

DeBus however, wasn't telling the entire story of his early retirement. Things had changed at NASA and they were very disturbing changes at that. The change would revamp the U.S. Space Program and demolish the dreams that DeBus and that early NASA generation had envisioned.

Apollo closes out and NASA realizes the future

Launching rockets was not the only thing NASA was doing. Apollo the moon mission was over, and the space race was done. It all seemed much less urgent and important in 1970. Things throughout the U.S. were very difficult, indeed.

The end of the Vietnam War was on the horizon with thousands of soldiers returning to what they hadn't a clue. Politics was looking at domestic problems like civil rights, the poor, housing issues, and a recession. It wasn't a pretty picture and the gilt on the rockets of space was no longer shining. In fact, many of the public didn't see the need for a space program anymore. Lyndon Baines Johnson, now president, was faced with a war that was hated and his plan for the "Great Society" had not done enough to ease the pain of the U.S. population. As LBJ left office, and Richard Nixon entered the presidency, it was known that Nixon didn't care much for space at all. NASA was now fighting for its own life. While there were plans for Mars, they were pipe dreams for NASA. With the cost in the billions, it wasn't about to happen. Much of the Apollo machinery was closed out section by section. The mighty Saturn V rocket was now just a museum piece. There was a hope that some of the lunar and Apollo material would be used for something like systematic lunar exploration, a space station or a robotic visit to Mars. While these concepts were adopted, they were hardly new or usable right now. How was NASA going to pitch the next great mission? What would be its next great mission? After the $25 billion spent on Apollo,

There wasn't much left for NASA or any other programs. By 1965, James Webb, then administrator for NASA, was trying to look forward but the future was seriously clouded.

By 1969, the Space Task Group was formed by President Nixon to outline a new space program. Nixon, in the end, did not act on the STG report that left NASA looking around for something to do. There was only 3 billion in the till for NASA in the 1971 FY budget. That meant that cutbacks and layoffs were coming. Along with that, morale for all NASA employees was in the trash. Trying to hold onto skill sets was also not easy since many of the workers would move on to some other type of work.

In the mid 1970s, NASA was looking at one program, SKYLAB. This was one of the first projects to come out of the Apollo program. While SKYLAB was not so expensive, used Apollo era hardware and was a reachable goal, it was somewhat of a hope. SKYLAB would give NASA at look at what a long term space station would need. The austere budget also allowed NASA to look at robotics and unmanned programs.

By 1973, while programs came and launched, the space shuttle looked like it would be a reality after all. While it was a hot controversial program, at least the decision to approve the shuttle and making KSC the launch center, made a turning point for NASA.

The shuttle idea went back to the early 1930s. The winged rockets created by Von Braun and Willy Ley were shown in a book they wrote called *"Exploration to Mars"* which was a best seller. The drawings and artwork were truly stunning. The idea of a reusable spacecraft that could bring crews and supplies to different planets was enticing. Reusability was the key to making space life a reality.

Another system ACES (Air collection and enrichment system) along with some other scramjet concepts were out there for design study. However, there were huge technical problems that required aerial refueling at Mach 6, something we haven't even accomplished in 2018 (at least outwardly, Lockheed is building a Mach 6 hypersonic demonstrator as of this writing). HIRES (Hypersonic In-Flight Refueling System) were almost given to the X-15 program but it was stopped before anything catastrophic could happen.

How would this new program affect the NASA centers? KSC certainly had it piece, Johnson Space Center would be known as Houston and had care of the astronauts, and mission control, Marshall was still the design capital and on it went to the various other smaller centers. During the building of the shuttle, Marshall certainly took the reins and was very jealous of its position, as was JSC with the training of the astronauts.

Even the U.S. Air Force, with the building of the Vandenberg launch complex, at Vandenberg AFB, in California as an additional launch site was acting protective of its turf. The Air Force wanted to control its own secret satellite launches due to the polar positioning of the Vandenberg site. On and on it went, as the infighting for place and position continued.

The NASA Michoud Facility and the External Tank

In 1973, the world of the space shuttle wasn't sunshine and roses. It was a tangle of weight, external tank designs, SRBs and many other things, but most especially money.

The complete design was to include the external tank and two solid rocket boosters (SRB). It showed in the drawings, yet nothing was done to build them. The components did not go into contract because they were the subject of phased procurement. This allowed the mature definition of one shuttle element before proceeding to the next. The weight of the Orbiter had to be locked down so that the external tank size could be controlled. Both of these major components had to be sized before anyone could design the SRB. The SRB and the ET had their own contracts for procurement and the competition that occurred in procuring them in 1973. The external tank was up for production at the Michoud Assembly facility in Louisiana. Michoud came into being in 1876 as a plantation grant for the French King Louis XV. Named for landowner Antoine Michoud, an eccentric junk dealer, he was never successful, yet tried to produce lumber from some of the swampland. Later on, WWII necessity built the plant to make and assemble Liberty ships for cargo. However, contract problems caused a shift in purpose and Michoud began to assemble C-46 cargo aircraft, only producing two aircraft in total.

By 1961, while Apollo was beginning its start up phase, the building went to NASA with over two million square feet to use. Due to lack of use, the building was a vermin infested, dirty and floodwater besieged wreck. Yet, with Apollo on the books, the Boeing Company, an Apollo contractor, cleaned up the mess and started construction on its Saturn V

Rocket stages. Of course, after Apollo, the shuttle program moved in. With that, the huge external tank would be a monumental configuration with a diameter of 27.5 feet, internal volume, six times larger than SKYLAB. It would be the largest vehicle ever built in Michoud.

The beauty of the Michoud complex allowed shipping by river from Michoud, Louisiana to Canaveral in Florida. At that time, the thought was to launch the shuttle from Vandenberg in California and this would have had the ET shipped through the Panama Canal, thus bringing it to California, quite a distance. By 1973, the cost for a tank went from $2.1 to 2.3 million, once production was well under way.

The tank along with the SRB looked a bit like the Titan IV. This wasn't lost on the Titan builder, Martin Marietta. The president of Martin Marietta met with NASA Administrator, James Fetcher on September 1972 and passed on a letter reminding Fletcher of his company's success. Tom Pownall, President of Martin Marietta, said in the letter, *"Since 1962, Martin Marietta had been involved in all phases of integrating and launching large solid rocket motors with large liquid tanks in the Titan III program. We were the first and until now the only contractor currently engaged in the type of integration which closely parallels the shuttle integration requirements. During these ten years, we have developed an in-depth understanding of the problems and solutions to successfully launch large SRMs with huge liquid tanks."* Pownall noted his firm was responsible for *"providing the design development qualifications and installation of electrical /structural-ordnance hardware needed to structurally attack and electrically interconnect the SRM to the Titan III and to laterally separate the SRM from the Titan III core during ascent."* Competition for the External Tank controls started in early April of 1973 as NASA issued a "Request for Proposal" (RFP) to interested bidders. Rockwell built the Saturn-II for Apollo and would have been a top-level competitor, but NASA wasn't ready to put all its eggs in one basket, and did not invite the bid since Rockwell already had the contract for the Orbiter.

McDonnell Douglas, builder of Saturn-IVB, received as invitation. Martin Marietta and Boeing, who had worked at the Michoud plant, had built the S-1C also got an invitation to bid. A RFP went to Chrysler, which operated as Werner Von Braun's manufacturing arm and constructed the Saturn 1 and 1B first stages at Michoud. The award went to Martin Marietta in mid August of 1973.

George Low, NASA Deputy Administrator said: *"The Source Evaluation Board presented us with a rather clear cut case for selecting Martin Marietta Company which we did. Technically, Martin Marietta was first and more successful company" "McDonald Douglas sent their cost both for design, development, test and engineering and for production work, which was considerably lower. Although we recognized that the Martin Marietta Company was partially "buying in" with their lowest costs we nevertheless strongly felt that in the end, Martin Marietta costs would be lower than those of any other contenders".*

McDonnell Douglas had a different kind of buy in. In that, they showed the weights below a reasonable value. Boeing's major weakness was their selection of General Dynamics as a partner in area where they really did not need this company. This presented as awkward marriage and probably would have caused major problems during the years. *"Chrysler was weak all around. In selecting Martin Marietta, we felt that they will undoubtedly do the best cost management job in that they clearly proposed doing this "the new way" (design to cost) the others in effect were doing business as usual."*

The solid rocket booster was little simpler to control than the external tank. The set up for the solid rocket boosters had a unique delivery system. They were shipped from their last stop, United Technology in California, via railroad. This was of course, after Thiokol had finished with them in Utah. The segments were round, drum shaped segments and already filled with the solid propellant that resembled rubber. Ten feet across, they were sorted in an arrival area and after inspection; they were moved to a ready storage facility. As the external tank was filled and further assembled in one of the four super bays of the VIB (Vertical Integration Building), this core vehicle stood on top of a platform that was forty -four feet across and fitted with wheeled trucks and rode on a set of railroad tracks.

The two solid boosters came together at the SMAB or the *Solid Motor Assembly Building* that was over two hundred feet tall. It was here that a crane lifted the sections and stacked them vertically with a nozzle assembly at the bottom and had five propellants filled segments with a nose section at the top to complete the vehicle. They stood some eighty -five feet tall.

In both the VIB and the SMAB, the wheeled platforms holding the main vehicle moved between the two buildings. The three hundred ton crane lifted and transported the completed solid boosters and put them along side the main vehicle. Still riding on the wheeled platform, the locomotive moved it from the SMAB to the Launch Complex. There were four major builders of solid rockets, Lockheed Propulsion, Thiokol Chemical, Aerojet General and United Technology. All these companies were exemplary in building and operating solid fuel rockets for the USAF and the Navy. All four of the companies knew the space shuttle requirements inside out. Early in 1972, NASA's Marshall Space Flight Center, which held all the responsibilities for the SRB, yet they knew little about the ins and outs of building it, gave contracts to the four companies to design at the 120 and 156 solid core models.

In July 1973, Marshall put out yet another RFP for the shuttle solid booster development, which again invited the same, four companies to give in their bids by late August. The Source Evaluation Board, which was the "grandma" overseeing the processes, which contained some 280 people who focused on the process, conducted their own analysis and design studies. In the basic structure for developing a solid rocket booster, usually ranged from building very big solid pieces from segments. All the companies had something to go on. Aerojet had a lot of experience with a 260 inch core and decided they were going to build "a single monolithic case", that had no joints or tang and clevis[6] accoutrements. It could be transported by water instead of railroad. Let's stop right here for a moment and explore Aerojet's design. This would have been the dream SRB, no moving parts, no O-rings, no putty. It was a simple, solid case that would have been large enough to give the shuttle what it needed in speed and strength. The Source Evaluation Board kicked this design to the curb. Their reasoning for it was, "The strength of the case was found inadequate for the pre-launch bending moment loads and was not designed with an adequate safety factor for water impact loads." What the Board was trying to imply was it couldn't land in the water with a parachute, after launch. When things have less moving parts, less can go wrong with them. That is a simple engineering fact. The Source Board didn't quite understand that or didn't want to.

The Source Board also felt there were other problems, and it gave it a fourth place in the competition. Next in the three remaining bids was Lockheed. Lockheed had a solid proposal. Yet, the Source Board felt that it couldn't control the costs. Thiokol was next but weak in technical areas. Its proposal "gave good reason to anticipate costs that would be both low and well controlled" United Technology had a bid that was mediocre and nothing to write home about. The strange thing here is

---

[6] Tang and Clevis— Mating section of field joints in the SRB motor. Clevis shaped like a "u" while the Tang is a straight pin. The Tang fits into clevis by sliding down the sides of the "U" shaped Clevis and connects two segments together. This joint the field joint with 2 rubber O rings in segment of SRB motor.

Thiokol's weakness in design yet just because it could say it felt it could "anticipate costs would be low and well controlled" it remained in the running?

To clarify the Source Evaluation Board, NASA Administrator, James Fletcher, headed it. That was an anomaly in itself. Why was the head of NASA on a board like this? In the Board's summary, it placed the choices for the SRB between Lockheed and Thiokol, claiming "the thoroughness and detail that Lockheed utilized in designing for low cost" was important. This case design was good, had ease of assembly, check out and no radical safety factors. The stress analyses, which were added to the proposal, showed Lockheed used the structural design of the case to meet load requirements. The nozzle was simple and used everyday materials for ease of cost.

Fletcher also brought out that it was "clearly focused on a highly reliable, low risk, conservative design, using all proven material. Why didn't Lockheed win the bid challenge? The problem had to do with management. Lockheed said they would use a quick set up of making a new facility and hire on more staffing. In the Source Evaluation Board's view was not a way to save money. Fletcher said that Lockheed's approach would end up with an early funding splurge, which was against the program's goals. According to Fletcher; "An anticipated relocation of personnel to the New Orleans area was not adequately planned or described." Lockheed showed that it already had staff and management issues with redundancy and over-control. The Lockheed project manager would have no backup and would end up being spread thin. Fletcher also brought out that Lockheed's key people had "no experience on a solid rocket motor program comparable in size or scope" that could relate to the shuttle program.

Next comes Thiokol. According to the Source Evaluation Board, Thiokol was ideal in management. Thiokol had no need for a new location, or to move anywhere. Of course, it would use Utah, its home base as its main facility. Thiokol was told that its proposal showed weakness. The nozzle used materials, *"not currently developed or characterized with attendant technical and program risk. The nozzle design was insufficient to meet required safety factors and could require a redesign" "Their design was intricate and would be a problem to manufacture"* the main problem with the design was *"in the area of case fabrication which could require extensive design."* Fletcher claimed he saw the "strength "in Thiokol's supervisors because they were "experienced and had worked as a team on Minuteman and Poseidon." Yet, what about the design? How good was the design? The project manager was considered very strong and was known for his excellence. His back up had important and successful engineering management roles in previous major motor programs and has an excellent reputation in the trade. Fletcher also said that Thiokol had *"depth of experience available within the company's operations department".* Here is the conundrum: do we choose a company because they have a weak design and good management? Alternatively, do we go ahead and take a great design, kick it under the bus and say that management was too weak to make the case and win the bid? Furthermore, one of the bidders had the "advantage" of coming from the home state of the person running the Source Evaluation Board. It was said that Fletcher's very tight connection to Utah business and the Mormon Church had no influence on his decision, but really how dumb do you have to be to see that there was a connection. No matter what Fletcher said, there was a connection.

Here we have the scores for the bid:

Lockheed 714 points

Thiokol 710 points

United Technology 710 points

Aerojet 655 points.

The Argument: Lockheed vs. Thiokol vs. GAO

Other board members were not too pleased with Thiokol's technical proposal, they weren't all that charged up on Lockheed management problems either. Since NASA had already chosen the solid boosters in deference to the liquid tanks early in 1972, the idea that the solids would be cheaper and easier to manage made the choice of Thiokol the way to go. If NASA had gone with Lockheed, it would have really canceled out their earlier decision.

George Low said that the Source board "decided that Lockheed had written an outstanding proposal, but that this did not mean that they could also produce a good rocket; and that the difference, which were purposely amplified by the Source Selection Board, were all in easily correctable areas." It seemed that the board was ready to hand off to Thiokol in what they felt were "easily correctable technical matters" since Thiokol's managers were the best in the business, or so they thought back then. The Board still felt that Lockheed had too many managerial obstacles to overcome, all the way down to lack of facility and inexperienced staff. The Board decided to go with Thiokol. James Fletcher, George Low and Howard McCurdy of NASA all agreed.

Fletcher: Did he or didn't he?

The whole scenario would all come back to haunt the Board in 1986 when Challenger blew up, due to those solid rocket boosters that were bought with the "experienced staff" and other things that were listed on the reasons why they were selected. Many things brought this home to roost on James Fletcher. Frank Moss, a Utah Senator was looked at as having placed a lot of pressure on Fletcher to make sure that Utah got the SRB business. There was the usual idea that Fletcher just had to maintain his status at the Mormon Church and the businesses he held in Utah. Although many felt that Fletcher was not susceptible to all this pressure, it's still hard to believe that some of it didn't rub off on him. No one can become a statue overnight and let everything just run off himself like water off a duck's back. When you look at the letter below, it's not hard to imagine that the fix was in, at least as far as Fletcher in a self-typed, uncorrected letter that was sent to Senator Moss was concerned:

"Dear Mr. Chairman:

I feel an obligation to respond to the numerous efforts made by your office of later to have this Agency, and, in particular, myself look with considerable favor at the placing of some of our business in your State. Not only would it be highly irregular to say the least, but might provoke the kinds of inquiries we are not prepared at this time to handle ... But the fact remains, Mr. Chairman, that my hands are tied for the time being. In my present position here at this particular Agency, it would be extremely difficult if not somewhat unethical for me to channel any more of our contracts towards your State without arousing further suspicion...

I would also like to call your attention to another matter along these same lines. One of your staff...went so far as to insinuate sometime ago that I had a moral, if not a spiritual obligation to acquiesce {sic} on some business issues previously raised by President Tanner. This person voiced an unthinkable opinion to the effect that my Church membership took precedent over my government responsibilities.

Knowing that you share similar sentiments with me in the clear separation of Church and State, I would like to request that you take this unpleasant matter under advisement with the individual in question and explain just how serious and unconscionable those inferences were.

But for right now I must pursue a course that, at least, seems to be equitable to all parties concerned. Sometimes substantive actions don't count as much as how others perceive them to be."

It is obvious that Fletcher was feeling the heat from his fellow statesmen in Utah. However, he still gave the contract to Thiokol. In a quote from Kim C. Gardner, Senator Moss's top aide, *"There's no question that one of the main reasons Thiokol got the award was because Senator Moss was chairman of the Aeronautical and Space Sciences Committee and Jim Fletcher was the Administrator of NASA. Moss's position gave us major clout in lobbying for it."* However, even Gardner turned tail on his statement about Moss and denied that the Senator had any influence at all. All of this made no issue to Thiokol, who was already sitting on a nice fat contract. Moss, who was looking at a reelection campaign in 1976, looked good because this brought jobs to Utah and of course, when Fletcher needed something from Moss, he was almost sure he would get the help he needed.

NASA finally decided and announced its winner to the bid on November 20, 1973. The other bidders were in shock. Lockheed was so upset it went to court. On December 5, Lockheed went to the GAO (Government Accounting Office) and told them they were going to court against NASA's decision. They looked like they had a solid case and filed suit on January 8, 1974, stating that NASA "arbitrarily and improperly increased" Lockheed's cost proposal and introduced unprecedented and improper design correction process in giving Thiokol the benefit of the doubt in its engineering work". Lockheed wasn't the only one fighting. United Technology was also angry and stated, *"Without record of achievement in the large solid motor field, it seemed incomprehensible"*. That was a straight up statement if there ever was one against the Thiokol approval by NASA.

The lawsuit put the SRB decision on hold until things were decided. Since NASA was not allowed to give Thiokol the entire contract, they could only allot them little parcels of money to work with. The GAO still had to figure out what it was going to do, and with the lawsuit in place, the money ran out fast. NASA was trying to push the GAO into making a decision. The GAO made the statement that it would make its decision by June 24, 1974. They were good for their word, and sent a ninety eight-page report to Lockheed agreeing with just about everything NASA wanted. There was one small problem however, that had to do with the production of one element of the solid propellant, that being ammonium perchlorate which was to serve as the oxidizer for the solid propellant.

Lockheed had planned to use Michoud plant in Louisiana and the Mississippi test facility for its test site. They also planned to send the segments down the river by barge. They would need tons of perchlorate (12,000 tons per year for a shuttle fleet in routine ops) Kerr-McGee was the manufacturer of the substance and the nation's biggest supplier. The other bidders, Aerojet, Thiokol and United, were going to meet their own needs using a local plant in Nevada. Lockheed was planning to build another plant along the Mississippi to make the perchlorate. It was going to cost money and that would put Lockheed back on its heels for some $122 million. This was one of the reasons Lockheed was tossed back in the pile as far as getting the bid for the SRB was concerned.

After this, the GAO looked at the actual production of the perchorlate and found that the United States was not exactly lacking in production of the chemical. Since it was clear that Lockheed was out of the running, whatever Thiokol needed was now freed up and that cut Thiokol's cost from $122 to $54 million. There were other issues floating around, but it pointed back to Thiokol as the one to get the bid, which they did in the final outcome. The final decision was made on June 26, 1974 and it was Thiokol hands down.

Politics yet again

More politics appeared, as there were Senators and Congressmen that were not happy with the outcome of the bid. Senator John Stennis, who was a member of the Space committee, really wanted the plant for Lockheed in Mississippi. The Senator saw much haziness when it came to the GAO report. Senator Russell Long of Louisiana also had some big questions regarding the GAO's decision.

A letter was written the GAO requesting that there would be some clarification, which could possibly help Lockheed. To make a long political story short, the letter didn't do much. James Fletcher spoke with Senator Stennis on the phone and told him that Thiokol had the contract and that was that. George Low, NASA's Deputy Administrator summed it up; *"Stennis apparently accepted that decision, but not very graciously."* Fletcher sent a letter to GAO[7] head, Elmer Staats that said, *"I have determined that there is not justification for reconsideration of the selection decision...I have directed that a letter of contract be awarded promptly to Thiokol Chemical Corporation."* The contracts were signed early the next morning.

Two years passed since Nixon announced his weak support for the shuttle program, and Rockwell got the contract for the Orbiter. Lockheed's fight with the GAO lasted some six months and that slowed down the SRB decision. To top that off, Watergate had just sprung a leak and Nixon was sinking fast. Not to be left on a ship that was foundering, the USAF decided they were going their own way when it came to any plans for a spacecraft.

The Orbiter

North American Rockwell was already into a detailed design phase for the Orbiter. The contract award was announced on July 26, 1972, which closed out the bidders that consisted of Lockheed, McDonnell Douglas and Grumman. The bidding was a complicated scenario because

---

[7] GAO General Accounting Office

it looked at not only the management issues but also the technical design. As NASA picked the proposals apart, their Selection Board broke down the details and scored the points for each proposal. Grumman and Rockwell were in front. Grumman had the high points when it came to technical concept. Rockwell beat Grumman out when it came to management, especially in showing it could support the lowest in cost out of everyone. Overall, it looked like Rockwell had Grumman beat. Rockwell wasted no time and started the ball rolling by hiring workers for their Space division. By 1975, Rockwell brought on over sixteen thousand workers for the new contract.

The final contract took some eight months to put together. NASA sent out its letter to go ahead with the design work. It didn't take Rockwell long to get it going, since they already had some four hundred engineers ready and waiting that had worked on the initial proposal.

Rockwell was already thinking that it was going to start farming out some of the shuttle sections instead of keeping the entire thing in the house to itself. For one thing, Rockwell had no idea how to handle the thermal tiles that were going to be needed for heat protection. Lockheed had done some work in this line and a contract went out for the tiles of the TPS (Thermal Protection System).

The shuttle program, as it evolved, was fraught with politics and this is where a lot of it started. Because of the nature of the design, many Senators and Congressman were happy that piecework for various shuttle parts would come into their states. Subcontracting became the word of the day. Rockwell decided that by issuing subcontracts, it would allow others to get a piece of the rather large Orbiter pie. Should Rockwell someday end up on the other side of the contract, the favor could be returned. George Low, Assistant Deputy NASA Administrator said: *"Even before the space shuttle selection was made, Fletcher, McCurdy (NASA Associate Administrator for Organization and Management), and*

I decided that it would be essential that a good share of the business go out to subcontractors to get the best possible talent in American industry and also to get the most widespread support for the continuing development of the shuttle...."It was always a good political move to "spread the wealth around". Low went on to say that; *"We discussed this immediately after the selection with Bob Anderson (president of Rockwell International), and with the leaders of the losing competitors. At the same time we began to receive pressure from the White House, through Jon Rose asking us to put an immediate subcontract on Long Island with Grumman. Our response was that we could not do this on a sole source basis, that is wasn't right to go to Grumman only, but that we could justify going to both Grumman and McDonnell Douglas to maintain their sound manned space flight capabilities. However, to do this we would need an extra $20 million in FY1973 expenditure limits..."*

The White House, on September 14, 1972, gave NASA the additional requested $20 million to put work out to subcontractors. The feeling was that the Aerospace industry needed to be fed to keep it healthy, so this was how it was going to happen. Rockwell subcontracted fifty- three percent of the work out to various companies. Rockwell Space Division took ownership of the Orbiter's nose and crew compartment, forward fuselage section and the aft fuselage. Rockwell was going to subcontract out the wing, mid-fuselage, vertical fin and the orbital maneuvering system. NASA was claiming that is was not subcontracting for subcontracting sake. They wanted to get the spacecraft designed faster. However there were problems with this idea, the Orbiter was badly overweight and that was a design disaster. It would end up costing more money to redesign and lose the weight. Initially, Rockwell had come up with an Orbiter weight of 253,000 pounds. Later on, it crept up to 277,500 pounds. When the total weight of the Orbiter, SRB and ET tank rose to 5,410,500 lbs, it was time for some serious slimming down.

The Abort Solid Rocket Motor

What could have been a defining factor for the space shuttle emergency escape system was dropped very early in the program. The Abort Solid Rocket Motors (ASRMs) were small format rockets, about thirty feet long and giving some 386,000 pounds of thrust. This would have been enough to take the Orbiter to a safe altitude and given the Orbiter enough thrust to make a safe landing at one of the emergency sites. The ASRM would stay with the Orbiter until the point where it wouldn't be needed and then it would detach. Yet, this couldn't be incorporated into the shuttle design. Why? Weight costs money. The ASRM could offset their own weight by thrust, allowing the shuttle to fly with less propellant, and it could also allow for shutting down the main engines after achieving a low earth orbit with the ASRMS helping to take it to a safer position. However, there were problems with the main solid boosters. It seemed that there would be no way to adjust thrust and with no vector control, there wasn't much to be done. Taking out the ASRMS would allow the design to be a much simpler way of doing it. There was still another problem. The SRBs were already in the plans for use as the escape rockets, yet they didn't have the necessary controls for thrust to make them viable. Once a Solid Rocket Motor is lit, there is no turning back, no way to control it. Hence, they would be useless as abort rockets. A NASA executive decided that after a study, the abort rockets weren't needed. Someone should have asked the Challenger crew about that. NASA engineers felt that the ASRMS were useful for only about the first 30 second of flight and a "gap" existed in that period during the first few seconds while on, and just above, the launch pad.

The ASRMS, created complexities in the overall vehicle design, operations techniques, abort modes selection logic, flight stability, mechanical systems, and software requirements, which made the system marginal at best. Failure modes existed in the ASRM system itself, which could cause loss of an otherwise nominal mission (e.g. failure to separate, inadvertent ignition) and possible loss of the Orbiter (e.g. uncontrollable CG (center of gravity) in the landing phase). With weight being a deciding factor, the thought of using the solid rocket motors for abort at low altitude looked to be reliable. The thought of using the SRM to gain necessary altitude and time to establish a flight trajectory back to the landing site for the Orbiter, was considered a safer situation than the time critical, complex and demanding flight control and systems requirements associated with the earlier abort solid rocket concept. It all sounded like it could work but would it?

In November of 1972, the ASRM was cut from the shuttle Orbiter design, which gave back 7,000 lbs and brought the weight of the entire shuttle structure to down by 99,000 lbs. This was thought to be a good idea, back then, but to be honest, how good an idea was it? Shouldn't there have been some sort of safety procedure built in from the start of the Shuttle/Orbiter design?

Even though it would have cost extra money, weight could have been reestablished to meet some sort of emergency contingency procedure. If this were done in the early design stages, instead of as an afterthought, we may have saved fourteen lives and possibly two orbiters. Due to money issues, politics and projects going to the lowest bidder, the shuttle didn't stand a chance of being the space vehicle it could have been. Yes, the shuttle was a mesmerizing machine, developed to what was then good standards, but let's face it, there was something missing. Could the word be less politics and more design integrity?

The USAF enters the picture and how

The United States Air Force was not happy about the institution of NASA. The USAF felt that it would have to protect its turf, that being the air and above, meaning space. The USAF had many of its own ideas regarding space and wanted to pursue them the way they wanted to. However, this was not to be. While the Shuttle was going to be a NASA program, the USAF by force of the Pentagon, was going to find itself a major user of the program. Nixon was almost out of the White House, thanks to Watergate, and the USAF was already dreaming about what they would want to do next.

In fact, if the USAF did not step in, the shuttle program might not have made it without the USAF as its main customer. The USAF intended to hold its turf and no one was going to take it from them. The USAF was very busy building rockets that would launch satellites for other agencies including themselves. Since they would become a part of the NASA program, the USAF started thinking about what it would need. In the end of the 1960s, the shuttle design was made to meet only what NASA thought it would need. NASA also felt that 25,000 pounds of thrust payload would be sufficient. However, after the USAF got a look at the payload bay, they decided that they would need something larger, something in the neighborhood of 65,000 lbs. The USAF wanted to send large satellites into orbit and what better way than to use the new shuttle. They put in their requirements for a payload bay of sixty feet by thirty feet. The USAF also decided that the shuttle would need a wing change, they wanted the Orbiter to carry a delta wing.

Even though, NASA and the USAF had to cooperate, there was always friction between the two houses. Vandenberg Air Force Base was the USAF's little piece of heaven on the California coast and they wanted to use it to launch its satellites into a polar orbit, something that couldn't be done at the Kennedy Space Center. The reasoning for this was that the polar orbits could give the USAF what they wanted to see in the USSR. The USAF rebuilt the base to support the shuttle. Vandenberg was a large enough base to suit the USAF and it was treated as the "west coast version" of the Kennedy Space Center, at least as far as the USAF was concerned. This is not to say that the USAF didn't already have many launch sites at Kennedy Space Center, and had been established there since the early days of Cape Canaveral. Remember at this time, both NASA and the USAF was still trying to decide which would be the fundamental base where the shuttle would call home. The USAF of course, felt that Vandenberg would be sufficient and when needed they could use Kennedy. However, NASA didn't have quite the same feelings. In 1970, the NASA Administrator, Thomas Paine decided that he would set up an advisory committee of both NASA and the USAF to take a look at all the alternatives at each of the many bases that they all were thinking about. That included White Sands, New Mexico, and Wendover, Utah which was adjoining Holloman AFB, and of course, Kennedy Space Center in Florida. There was also the thought of how and where the different components needed for the Shuttle would be delivered. This needed to be put into the equation on choosing what and where would be the home base.

By December 1970, Congressman Olin Teague, Chairman of the Senate Subcommittee on Manned Space Flight made a statement that: "Unless I am convinced that NASA is making maximum use of the existing facilities, I intend to oppose any money for the shuttle in every way, shape or form". That statement pretty much put the ball in NASA's court for a Florida base. NASA had already started work on its Florida facilities so that the shuttle would have everything it needed as far as launch abilities were concerned. Yet, since there were constant changes to the shuttle design, the direction of where to "hold court" also changed. It was a matter of how and where the different components were delivered. Easily, Florida seemed to be the place to go. The external tank had to come by water; it was too big and bulky to go any other way than by barge. As for the SRM, it had to come in segments. Railway and barge was the only safe, logical way to accomplish that. In the long run, the

Solid Rocket Boosters had to land in water after the launch to allow pick up and delivery back to base. The Atlantic Ocean was the only way to go. However, the USAF did not give up, and continued to rebuild its site so that the SRB's could be recovered in the Gulf of Mexico. With all this bickering back and forth, the final announcement was made in April of 1972, when George Low, Deputy NASA Administrator said that Kennedy Space Center was going to be the Homeport for the Shuttle, with Vandenberg on the west coast, as an alternative launch site. Vandenberg did have some issues like not having a long enough runway and would have to be lengthened to fifteen thousand feet. The good thing was that Edwards Air Force Base wasn't too far behind Vandenberg geologically and could be used for any abort landing or emergency landing.

By 1975, the USAF put some $650 million aside to refurbish the Vandenberg in to shuttle homeport. NASA was looking for a launch by 1979 and that was extreme at best. Now that the USAF had it say as far as the shuttle configuration of the payload bay and the fact they would have Vandenberg as a shuttle homeport albeit secondary, life proceeded on and construction of the shuttle continued. While the USAF looked into the future, it was also looking over its shoulder at short-term needs. It was busy upgrading its Titan rocket.

NASA too, was looking over its shoulder and realizing that it too, had competition. Just because they had the shuttle, didn't mean that everyone would be flocking to their door to sign on. There was still the USAF to look at and that is not to even mention the Ariane[8] missiles over in Europe that were flying successfully. While all this was going on, the shuttle was falling behind schedule.... as usual.

---

[8] The Ariane missiles were built and launched by the European Space Agency. It was a heavy lift vehicle capable to deliver a payload to a geostationary orbit or a low earth orbit.

Delta wing or ?

The Delta wing was favored by the USAF in that would have allowed the Orbiter to do more cross range flight on reentry than the straight wing would have. Cross range was the key to the USAF using Vandenberg for a one time turn over in a polar orbit, or to grab a satellite and come back. However, the delta wing, being larger and heavier would need more protection from heat, which didn't help the weight problem at all. There were ways to look at a delta wing that could help with weight issues. Rockwell was determined to find it, via many hours of wind tunnel tests. Rockwell still had to satisfy the USAF prerequisite of cross range for eleven hundred nautical miles and get low speed lift for performance at approach and landing. There was also the question of landing which would add to how the leading edge was designed. The maximum angle of attack for the shuttle on landing was seventeen degrees. Finally, the developed wing was a "50 degree blended delta". This wing had straight trailing edges with elevons, and a curving leading edge that blended into the fuselage. This wing went through the *Program Requirements Review* in November of 1972 and showed that further weight still needed to be dropped. Both Rockwell and NASA brought the weight down by choosing to go with a higher landing speed of 165 knots and an angle of attack at fifteen degrees instead of seventeen. This managed to save weight by shortening the landing gear, allowing the dry weight for the shuttle to go from 170, 000 pounds to 150,000 pounds.

Changes

Many changes in the shuttle brought with it a large quantity of control that in shuttle design was only at a bare minimum. There were still things that weren't locked down yet, like the solid rocket booster which were designed for 156 inches diameter, allowing railroad transport. However, Rockwell wanted these boosters to be 150 feet long. Overall, the fact that the shuttle had reached a new lower weight, made everything more compact. The very urgent problem of cost per flight now came to $10 million, which is what NASA had decided on earlier.

Rockwell was finally able to decide its subcontractor contracts by March 1973 and declared its winners. This was turned over to the Johnson Space Center in Houston for their recommendations and on to NASA headquarters on March 26, 1972 to be approved by James Fletcher, NASA Administrator three days later. Grumman Corporation got the good news that it would be building the wing for the orbiter for $40 million. McDonnell Douglas, that had a huge layoff problem, was saved by a $50 million sub contract offer for the orbital maneuvering system. Fairchild Republic Division got a $13 million for the vertical fin and Convair of General Dynamics received a $40 million subcontract for the mid fuselage that included the payload bay. Rockwell went on to do its part building the nose, crew cabin, forward and aft fuselage. The blessing for Rockwell was that its Rocketdyne division was on contract for the main engine for the Orbiter. However, it wasn't all that simple. Rocketdyne won the award for the SSME[9] but Pratt and Whitney, another contender was not quite settled on the fact. Pratt and Whitney protested that NASA specified that they wanted an engine bell configuration, nozzle bell as opposed to Rocketdyne's aerodynamic spike[10] which they promoted in the 1960s. Pratt and Whitney was looking at a bell nozzle for the SSME. Pratt and Whitney was not happy and in fact did protest the award. Even after nine months or protest, the award was settled on

---

[9] SSME Space Shuttle Main Engine

[10] This aerodynamic spike was in the line of the truncated nozzle: its packaged and derived the nozzle from the expansion of the gases where the nozzle wall was formed by additional gases that came out in the cooling system. It was a high-performance engine. It did save 10-12 feet and the weight that went with it (J.R.Thompson-design development test and operations of the SSME Director of the Marshall Spaceflight Center NASA)

Rocketdyne. However, there were many of the NASA engineers who worked on the Apollo project that thought Pratt and Whitney really had the edge on the SSME and many felt it really was Pratt and Whitney's turn to have a shot at the engines. Sadly, it didn't come out that way. It was a great proposal but Rocketdyne somehow had the edge. While Rocketdyne did get the engine, they really weren't up to snuff as they say. According to J.R. Thompson who was the Director of the Marshall SpaceFlight Center, Thompson said that he felt Rocketdyne "took a big sigh of relief" when they won the contract, but there were some problems. At the Rocketydyne test facility in Santa Susanna, California, things were not what you could call shipshape. In fact, they were in very poor shape, management and facility wise. The place wasn't kept up to the standards needed and it really got Rocketdyne into some trouble. NASA had one of their senior engineers by the name of Norm Reuel go in and straighten the place out which he did. Again this just shows how things were not always book perfect for the shuttle, not even in the early days when it really counted to get it right the first time.

Meanwhile, the list of sub-contractors grew on into 1975, which showed that eighteen projects were under subcontract. The contracts spread throughout the country with some thirty- four thousand workers nationwide working on some part of the space shuttle program. By 1977, the total had hit some forty- seven thousand.

Managing a Behemoth

The Space Shuttle Project was a huge undertaking. The NASA centers had some say in what went on with the program. Marshall Space Flight Center and Johnson Space Flight Center were among the two biggest. Actually, Marshall became a kingdom until its own self. With Werner von Braun as its director, and the larger missiles and rockets he had put together, Marshall had its own way of doing things and no one interfered. Both MSFC and MSC were the two highest authorities, next to NASA headquarters in Washington D.C.

In addition, right here is where some of the later problems started. According to NASA policy, each has to perform as the "lead center"

That was playing two ends against the middle and that spells disaster no matter what industry you are in. You will always have conflict and divisiveness whenever put into this situation. It's inevitable. That is exactly how it turned out.

During 1969, there was a shuttle concept that was agreed on by all. It was a two-stage, fully reusable configuration with each stage piloted, winged and powered by a liquid fueled Space Shuttle main engine. In September and October of 1969, managers from both centers came to a joint agreement that extended the two-part approach to the shuttle itself. Robert Gilruth and Von Braun both agreed that the development of the two shuttle stages that being the Orbiter and the booster were handled by separate contractors. The separate centers could manage each contract with Marshall in charge of the booster and Johnson in charge of the orbiter. NASA headquarters would control the *Office of Manned Space Flight*. By 1970, Dale Myers[11] arranged the shuttle program office with Charles Donlan[12] as the manager. Donlan was the first to say that Johnson Center should not only hold responsibility for the Orbiter but also be the lead center for the entire shuttle program essentially. Here began the turf wars inherent in the shuttle program. This started out small but did grow to the mess that later brought down two shuttles and killed fourteen astronauts.

---

[11] Dale Myers- Associate Administrator Manned Space Flight replacing George Mueller in 1970
[12] Charles Donlan—First director for the Space Shuttle program 1970-1973

According to Leroy Day[13], Donlan's deputy also recalled a deep resolve to condense the staffs in Washington D. C. by putting more responsibility at Johnson Space Center: "There was a ground rule laid down that said we didn't—we weren't going to have any support contractors. Therefore, whatever we had in Washington, at least the program office there would be quite small; we would not be large." The comparison is that, at its peak, the Apollo program office in Washington, including all the support contractors, was about five hundred people, if you can imagine that. This included Bell-Com, General Electric and a large number of Boeing people. We were not going to have any support contractors; we were going to have a lean program office in Washington. We would rely on the centers." That was the first mistake. The idea was that this would allow all to draw on experience that the centers has managed to gain during the Apollo years, which certified them to take on more responsibility in the shuttle program. That wasn't the end of the argument. The second part of the fight had to do with the fact that the shuttle was a flight vehicle and Donlan had this to say: *"My argument for the management structure was this: the Shuttle is a system. It's a system that is composed of these elements: the orbiter, the tanks, the boosters, and everything else, and ought to be designed not only as a system but managed as a system. This means that, at all times, the status of the orbiter should be the principal consideration, i.e. what happens to it. It's the element of the system that's going to pay off. Anything that affects its weight or its performance has to be weighed very carefully. That means if, for whatever reason, the Orbiter's size is to change and requires a change in the tank, that would have to be done, and so on, right through the system. So how do you manage something like that? My concept was to have prime management where the Orbiter was the prime element;*

---

[13] Leroy Day—Deputy Director of the Space Shuttle program 1970

that the boosters and the solids were subordinate elements. I structured a way in which Johnson {Space Center}, as the orbiter manager, would retain the prime field management for the system. And that Marshall would be delegated to carry out the management of the tank and solids, would indeed do that independently, but would report through the Orbiter office of the shuttle program Office, and similarly for KSA {Kennedy Space Center}, where all the requirements for launch facilities were being delegated. That created quite a furor, initially. Finally, Dale bought the idea. Moreover, he and I went over and sold George low on it. Then, Dale and I went and met with the directors and deputy directors of each of the three centers and discussed the proposed management plan with them."

NASA's framework goes something like this: both Kennedy Space Center and Marshall Space Flight Center are equals with Johnson Center but still remain under the Office of Manned Space Flight. Except under Donlan's plan, the lead center methodology, the shuttle project office at each of the other three centers was not to report to the main office in Washington D. C. The center office would be at Johnson Space Flight Center and this would allow JSC to have a two level management one for just the orbiter and another for the whole shuttle program. Johnson would hold the responsibility and would manage the entire shuttle issue along with the booster as well as handling all of the launch facilites. That sounds totalitarian, doesn't it?

It was also interesting that Marshall didn't want JSC to gain anything more than they had and vice versa, and there was no way that Houston was going to be second class citizen to Marshall. Now it was time to bring in the politicos. Senator John Sparkman of Alabama, the home state of Marshall got involved and the final management plan was announced on June 10, 1971.

Marshall got responsibility for the booster and the main engine that was to serve the orbiter. JSC was the lead center for the shuttle entirely. Marshall felt that JSC "would have "program management responsibility for program control, overall systems engineering and systems integration and overall responsibility and authority for definition of those elements of the total system which interact with other elements". JSC still held the responsibility for the Orbiter. Marshall sure was pleased. All of this information came out in the *NASA Management Instruction NMI8020.18* that was printed in July of 1971.

Other things were part of the NASA confusion. Little things like words for example. Project and Program meant different things. Program in NASA speak meant "coordinated major undertaking, within which two or more projects form the building blocks. "The shuttle as a "program" meant that it's solid and liquid rockets, tanks and the orbiter and launch facilities, all were "projects". Hence, this is the start of the verbalism and the ingrained confusion, which began within the shuttle program.

LEVELS of Management

This is a very interesting part of the NASA life style when it comes to the shuttle program. The different levels of management are most likely the most confusing of the entire set up when it comes to how NASA controlled the shuttle program. We will go through each Level and show where it comes out in the shuttle program.

Level I - resides at Washington D.C. headquarters. It gives the program direction with the Office of Manned Space Flight. This level deals with budget, schedule and performance requirements, allotting funding to NASA field centers. Level I decisions cover the solid rocket boosters and the payload bay with the USAF dimensions of 60 feet by 15 feet. The main executive would be the Associate Administrator for Manned

Space Flight, who heads the Office of Manned Space Flight and the Space Shuttle program director.

Level II - holds the overall program management within guidelines that are developed at Level I. It deals with integration of both the systems and project managements, to ensure that all major components of the shuttle go ahead when needed. Example, when the main engine is ready and when the orbiter isn't ready, everything goes off schedule. Level II management lives at Johnson Space Center in Houston. Its top executive position is the Space Shuttle Program Manager. All project managers at Level III report to this office.

Level III - The Level III offices reside at both Kennedy Space Center and Johnson Space Center. Johnson Space Center had a project office for the Orbiter with a project manager. Marshall also held a Level III responsibility for the major propulsion systems like the main engine, SRB, and the large external tank. Kennedy's office maintains the shuttle launch and landing responsibilities. In 1971, the hopes for this system are high. Johnson Space Center hoped that the JSC Level II maintained its even strain.

Houston was now approving its own facility needs and chucking anything that Marshall proposed. This was only the beginning of the backbiting that will show just how convoluted this whole process was. Deputy Administrator, George Low and Associate Administrator, Rocco Petrone received a letter from John Yardley, Assistant Administrator for Manned Spaceflight, who was in charge of the Office of Manned Space Flight in 1974. He wrote about discussions with Chris Kraft, Johnson Space Center director; "Management surveillance and evaluation of sister center activities are very weak. R.L. Thompson, doesn't believe it should be beefed up, but, rather feels that once the technical direction has been given, it's clearly the other center's job to produce and he feels little, if any, responsibility if problems arise." Chris Kraft, a NASA veteran from the old Mercury days, looked at the message and felt that Johnson Space Center could not direct Marshall Space Flight Center as "a practical matter". The problem was made more complicated with the consequence that both centers have different management styles. Actually, all the centers had different management styles.

Here we have the core of the problem with what goes on at NASA internally with communications. If one center doesn't speak or react to situations much the same way that the rest of the centers do, you have a beginning of the proverbial "Tower of Babel". That meaning, while at one time all spoke the same language, NASA devolved into this miasma of infighting, and political back biting for the project as a whole. Each section did what it felt was necessary for the project, but they distrusted each other and this really hurt the entire project.

Yardley went on to say: "Lead Center allocation of funds has a number of problems…the sister centers are always suspicious of the objectivity of the lead center".

"The different management styles created problems here, also. In addition, the matter of fund recommendations for program support put the program managers in a difficult spot between the center director and the program director. That required a detailed headquarters review to assure objectivity." Yardley went on to say, "no major organizational changes need be made, but that headquarters be reorganized to perform the missing management surveillance function and to play a more in depth role in the fiscal area. This essentially means the lead center becomes the lead technical center which really more closely describes the role they play today." In essence, what this says is that each center is jealous of its own funding and its own backyard. All of this brought about the return, partially, to something of what the Apollo program had, continuity and peace among centers.

It also brought about the facts that the lead centers Marshall Space Flight Center and Kennedy Space Center report to the Office of Manned Space Flight. They also now had the authority over their Level III project, with their relation to the Office of Manned Spaceflight, which now allowed them to bypass Johnson Space Center in Houston, totally. Washington headquarters also needed to reinforce itself since it worked in Level II and III. Responsibilities remained in one piece at the centers, while Office of Manned Space Flight took a heavier role in the everyday management and budget recommendations from Level II at Johnson Space Center.

Level IV Prime Contractors

The prime contractors were so far down in the chain of command, they merely carried out whatever job was handed to them from NASA. However, consider what goes on here. With each center speaking another "language" and being filtered through whichever center is in the lead, how much confusion are you seriously left with?

How long did it really take before some sort of dialogue was in place between NASA and its contractors? When would the shuttle finally get to the point that it looked like something? The external tank and the SRB were still being worked and of course, the USAF wanted to be very involved with Vandenberg. Things were coming together somewhat, but what of the NASA management?

The Orbiter Tastes the Air for the First Time

By December of 1975, the shuttle was under construction. In a trade magazine, a photo showing the spacecraft in final assembly in a hanger was the feature photo.

The title read "Space Shuttle 1976". This special creature was known as "OV-101". While this orbiter was never to see the realms of space, she was the first and the only atmospheric shuttle that was created. Her main goal was to test the glide features of the Space Shuttle. The timing couldn't have been better for the shuttle program, it was the shot in the arm it needed to show off what was coming.

The story about how *Enterprise* got her name is legend. While she was slated to hold another name, the "Star Trek™" TV show, "trekkie" fans saw their chance to put the stamp of TV history on OV-101. The name "ENTERPRISE" was petitioned in by many, many of the "trek-kies" and they won the day. Besides, NASA loved the public relations boost it gave them. Enterprise rolled out, September 17, 1976 and made her debut to the "STAR TREK™" theme with the principal StarTrek performers in attendance. It was a proud day for NASA, because it showed people were actually interested in space and they had the proto-type for the Orbiter that would fly in space. Enterprise was set on the back of a specially modified Boeing 747 that for forever after was called the "SCA" (Shuttle Carrier Aircraft). This Boeing 747 made it possible for NASA to test the glide characteristics and carrying the shuttle home from a landing in California. The Boeing aircraft's number was N905NA. The SCA was a complex system. It incorporated a mate/de-mate device that allowed the shuttles to be mated on the back of the 747 and de-mated from the back, while traveling between landing sites. Most new prototype aircraft have to make their way by taking off and landing. That was the proof of the pudding. However, in this case, the Enterprise needed an assist. Her mission was not only flight test, but to allow the pilots to try the glide landing that would be the hallmark of the Space Shuttle. An engine-powered orbiter would have been the dream, but it wasn't working out like that. The magic word was WEIGHT. Weight was the demon that plagued the Shuttle from inception. Hence, the 1970-71 configurations of the Shuttle showed a reusable planform with a first stage that was winged. The idea was it was supposed to fall back into the atmosphere from several hundred miles to the launch site. It had no other options but to get back to the launch site under its own power. The concept showed some twelve jet engines in the assist. Weight was a definite factor and it was not going to happen, at least not in this configuration.

June 1970, General Electric received a study contract along with Pratt and Whitney for a nine-month study of shuttle air breathing engines that used hydrogen. Turbofan engines included Pratt and Whitney's F-401 engine and GE's F-101 engine which had been used in the B-1 Lancer bomber for the USAF were looked at. With it mounted on the shuttle orbiter, the weight was an undoable 200,000 pounds. However, the hydrogen fuel would cut the weight some 2500 pounds. Yet, it still was not going to be enough. The testing of aircraft in unpowered landings was going ahead at Edwards Air Force Base, of which NASA's Flight Research Center was part. The Flight Research Center and Edwards had a unique relationship, NASA, as always, was above the fray and operated on its own. It wanted nothing to do with the Air Force and likewise the Air Force with NASA. With the X-15 in use at the time, there were many new concepts that were coming into play. The catch phrase for the day was "energy management". The plan went like this; from 35,000 feet, the "high" key point a pilot would make a 180-degree turn and head for the dry lake bed below. By the use of speed brakes and turns, the pilot could slow himself down and usually the X-15 would land safely. Much the same process would be used for the Shuttle, since the Shuttle was a lifting body. However, this showed that sink rates would be applicable to the shuttle un-powered landings.

The Space Shuttle was supposed to be intrinsic to a two-part system. That means the space shuttle and the space station were to be together as a means to an end for each other. The Shuttle's basic policy was that it could carry some 25,000 pounds of space goods back and forth to the space station that is when it was built. Because of that stipulation, the shuttle almost undid itself by locking itself into this configuration. When it came to Congress voting on appropriations and budgets, the votes were tight and close. NASA needed all the help it could get, it couldn't stand to lose anyone or anything if they were going to hold onto the Shuttle. The problem here was the Air Force was the one who was holding the reins, everything would be fine…. if the USAF agreed. NASA needed to make sure the shuttle that NASA conceived met the requirements of the USAF. That was a big chunk to swallow for the Agency, not only as a hit to its already growing ego, but politically.

As we already described, the USAF had its set of requirements for the shuttle and the main sticking point had to do with the payload bay being able to carry the huge satellites that it needed. They needed 40,000 lbs for a polar orbit, which meant 65,000 lbs for anything eastward, launching from Cape Canaveral. The only way to work it out would be to dump the jet engines and that would give the shuttle the weight advantage that she needed to meet the USAF's requirements.

The bottom line came to this; in 1971-72 as the shuttle's configuration changed every other day; the concept for the fully reusable winged first stage went the way of the wind. It was at this point that the SRBs appeared. Since there would no longer be any internally carried propellant, the shuttle weight decreased. However, the jet engines were still in the ever-changing picture. What the jet engines or the ABES (Air breathing engine subsystem) provided for was to allow the shuttle to make a safe, controlled landing during atmospheric or landing flight.

It would also allow the shuttle to remain airborne for any type of wait during a reentry phase. The shuttle would also be able to ferry itself to different landing sites.

However, since the 65,000 lb. payload and the jet engines would add up the weight, eventually the jet engines would be dropped. The loiter idea was something that the pilots really weren't charged up about because they felt it would give that false sense of security (Like anything that would aid in securing the shuttle's safe operation would be something unneeded). Un-powered flight, thanks to the X-15's magnificent performance, secured the shuttle's requirement for un-powered glide.

By July of 1972, North American Rockwell's design going to carry two mission engines, which deployed from the payload bay along with two more engines to be used in ferrying, attached to the struts. By December, this scenario was tossed and the engines were put on the shuttle's undersurface, and taken off for flight to orbit. The problem was this engine theory was adding up and both sides, both the USAF and NASA weren't explicitly happy with the whole thing.

The Shuttle was supposed to fly from one coast to the other. To have emergency fields throughout the USA would have been extremely expensive. The decision was made to drop all the engines and the shuttle would now be place aboard a Boeing 747 for ferrying that could cross the USA with no fuel issues.

C-5A or 747?

Before Enterprise could go anywhere there had to be an aircraft that would carry her. The first thoughts for the new Shuttle Aircraft Carrier went to the Lockheed C-5 Galaxy transport that the USAF was flying. The USAF would have further loved to annoy NASA by supplying the lifting craft for the shuttle. The good thing was that since the USAF was flying it, it was already in the inventory and there wouldn't have to be a new design brought on board. However, even though the Lockheed C-5 Galaxy was the possible choice for the mission by NASA, it was dumped in favor of the 747. One of the reasons had to do with the 747 and the low wing design as opposed to the Galaxy's high wing planform. The fact that the USAF owned the C-5 was another issue. The USAF would own the C-5 outright. If NASA went with the 747, NASA would own it outright, and that would make NASA feel a little bit better by not being so dependent on the USAF.

One of the main concerns for the shuttle and the SCA would be that the shuttle would be able to make a clean getaway from the carrier aircraft without hitting the tail of the 747. There were some 600 hours of wind tunnel testing between Lockheed, Boeing and McDonald Douglas. There was also a computer study with some 2500 computer simulations to make sure that the shuttle and the SCA could separate without a disaster.

Since both the C-5A and the 747 were easily known quantities, the extra thought went to the huge C-97 (now at the Nat'l Museum of the USAF), which the USAF looked at for cockpit configurations. The choice of the 747 was almost easy considering the C-5A had a T-tail and a horizontal stabilizer on the vertical fin. This configuration would cause the aircraft to pitch forward toward the Orbiter during separation. The C-5A pilot could stop a collision with the tail using a larger and carefully timed movement on controls, but not done correctly and it would cause a disaster by shearing off the horizontal stabilizer. The C-5A was looking less and less a candidate.

Yet, the 747 was a bit safer. There was no T-shaped tail to contend with, and only a conventional configuration that was mounted on the horizontal stabilizer below the Orbiter. Below that, mounted to the fuselage, the vertical was left standing alone. The aerodynamics were more attune to air launch. According to John Yardley, assistant administrator for manned spaceflight, "The inherent characteristics of the 747 is to pull away from the Orbiter thereby aiding separation. A pilot would not need to take the sudden evasive action to prevent collision. If the Orbiter did collide with the fin (horizontal stabilizer) being below and out of the way, consequences would not be catastrophic. The 747 could lose a large portion of this fin and still get back safe. The 747 had some other advantages. With the Orbiter, it had a range of 2,340 nautical miles, enough to cross the country without refueling. The C-5A needed in-flight refueling and had less range.

The USAF felt it necessary to develop the experience by flying a C-5A with a dummy orbiter. This was not necessary with the 747. The 747 safety factor had to do with the presence of the orbiter and that was most destabilizing while mated, allowing flights of this type a reduction in stability prior to flight with an air launch. This was not possible with the C-5A because its greatest destabilization occurred just after separation. This difference resulted from the dissimilarity between the two aircraft tail shapes.

The 747 could use a shorter runway than the C-5A if an engine failure occurred. The 747 count mount engines of greater power to increase the air launch altitude. Carrier aircraft were expected to see extensive action when the plan was for the shuttle to fly sixty times a year, way back when in the planning stages. The 747 was structured for a 60,000-hour life while the lifetime of a C-5A was no more than 12,000 hours. Commercial 747's flew everyday where the C-5A flew less frequently.

747 Gets the Job

The decision to take the 747 came from within the Marshall Spaceflight Center in Houston, Texas. In May 1974, the center director, the renowned Chris Kraft, wrote to William Scheider in Washington D. C., the acting Associate Administrator for Manned Space Flight, *"Dear Bill, This letter requests authorization for NASA Johnson Space Center to purchase a Boeing 747 aircraft."* On June 13, 1974, the shuttle program office comparing the C-5A to the 747 and recommended a 747. A committee concurred the next day; George Low saw the briefing and gave the approval request to Kraft's request of a month earlier. The buy went fast with NASA paying some $15.6 million for a used Boeing 747-123 from American Airlines. NASA did not name the aircraft but just put the new registration number on her, officially adding her to the space program. The first of the 747s chosen was tail number N905NA and still had the American Airlines piping strips on her side when she was testing the Enterprise during the 1970s.

The aircraft was taken on board by NASA in 1974 and was used in trailing wake vortex research as part of another study the NASA Dryden was running along with the Shuttle testing. In August 1974, NASA took ownership of its 747 and contributed to aviation safety by conducting thirty wake vortex tests in research flights. Wake vortices are narrow regions of extremely severe turbulence that trail for miles behind the aircraft wingtips. They are stronger when the aircraft is larger. During the test flight, a T-38 flew behind a 747, testing for wing vortex turbulence. She rolled violently behind the big 747. Safe separation between such aircraft when landing would be three times as much in that distance. Due to the fact the orbiter was mounted in front of the vertical fin, it reduced the fin's effectiveness. To restore diminished stability of the 747, designers made rectangular fins, 10 feet x 20 feet to fit on the ends of the horizontal stabilizers. The wind tunnel tests at the Boeing's main plant verified that they would be useful.

Boeing won a $30 million contract from Rockwell to carry through the 747's physical modifications. The work took place at the 747's production facilities near Everett, Washington, between August and December of 1976. There were significant structural modifications. The commercial 747 had been built to carry passengers or air cargo on a strong deck in the fuselage. The fuselage withstood internal cabin pressure and external aerodynamic forces for passengers, not built to carry the 150,000 pounds of the Orbiter on her back. The fuselage needed additional bulkheads for extra strength. The 747 used "stressed skin" design. Loads, weights and stresses were carried not only by the internal framework, but carried in parts of the skin, which served as a major structural element on its own right. This skin was reinforced in a critical area with overlays of sheet aluminum being riveted into place.

The SCA was modified by Boeing to carry the first class seats up front for NASA passengers, while the main cabin was stripped and new mounting struts were added to strengthen the fuselage. The 747's vertical stabilizers were added to the tail to help with the stability when the orbiter was mated to the aircraft. Engines and avionics were also upgraded and an escape tunnel added. This flight crew escape tunnel was removed later on in the program after the approach and landing tests (ALT) because of the issues that were possible with engine ingestion of an escaping crewmember. Not a pleasant thought at all!

With the additional weight of the Orbiter on her back, the 747 used significant fuel and had other altitude problems to contend with.

Additional fittings were placed on top of the fuselage supported the orbiter itself. These struts were mounted in areas that matched the socket fittings that mated the orbiter to the external tank. Forward struts had the shape of an inverted V poking into the underside of the Orbiter and came in two varieties. Those intended for approach and landing tests could telescope in length to raise the nose of Enterprise for easy separation, the second version for use of ferry missions were fixed in length.

Rear Orbiter supports were mounted on top of fuselage. Structural modifications added some 11,500 lbs to empty weight of NASA's 747. It was given a stronger and more capable rudder control. Extra weight demanded more power from aircraft engines, which also were improved. Cockpit controls needed for air launch and ferry missions were included. These included a sideslip indicator because the 747 had a tendency to you when matter with the orbiter. New equipment included telemetry and frequencies close to one gigahertz and transponders and frequencies five times higher.

Under separate contract, the JT-9D engines from Pratt and Whitney were refitted to boost the maximum thrust from 43,500 to 46,950 lbs. Modified for water injection, this sprayed water into the engines hot internal airflow during takeoff thrust for use in ferry missions when the 747 would carry both the Orbiter and full load. Work was completed on December of 1976. The 747 flew from Everett, Washington to Seattle and to the Boeing field for a flight test program that ran through the Christmas holiday. In January of 1977, Boeing turned the aircraft over to Rockwell, which flew it to Edwards AFB and the Orbiter, Enterprise, thus ending several years of preparation.

Roll out

Structural assembly of the Orbiter didn't take place at Edwards but at the Air Force Plant 42 in Palmdale, California. The work that was started in June 1974 in time for major components to arrive from Rockwell subcontractors and from that firm's Space Division.

It appeared appropriate to demonstrate through flight test that an aircraft resembling the orbiter could execute an un-powered approach and land with accuracy on a concrete runway. This wasn't done with lifting bodies, they had to land on marked airstrips within a dry lakebed, which offered more room. The SCA now had to prove itself by flying with additional weight and drag using the fuel and altitude issues that it would face to solve the problem. The range was reduced to 1,000 nautical miles and the SCA would now need to stop frequently to refuel while going across country with the Orbiter in tow. Even without the Orbiter, the SCA would need to carry extra ballast to balance itself out. The SCA would now need an entire work week and over 170 people to get the Orbiter and the SCA ready for a flight.

Refueling in the air

The USAF wanted to be able to have the SCA refuel in the air. It was a nice dream, but not to be. During the formation part of the flight, minor cracks were appearing on the tailfin of N905NA. Not having a sufficient answer why this was occurring, the USAF backed off the aerial refueling plan because no one really wanted to add any extra risk to the issue.

The first flights of Enterprise With the atmospheric orbiter now mounted to the back of the SCA or Shuttle Carrier Aircraft, the SCA now had to run its taxi tests, just as it would have as a new aircraft. The mated orbiter and 747 ran down the runway at Edwards AFB. These tests would verify whether the mated pair would be able to take off safely under the 747's power from her own engines. It would eventually set the takeoff parameters; assess the directional stability, control and elevator ability during takeoff. The 747 had to evaluate its ability in pitch, thrust reverse and brakes with airframe buffeting.

The Tests

Enterprise and the SCA had three taxi tests before getting to the meat of the program. These tests occurred all on the same day, February 15, 1977. The first of the taxi tests reached some 76 knots, which was under takeoff speed, at which point, the test pilot Fitzhugh Fulton (of the XB-70 program) hit the reverse thrusters and then the brakes.

On inspection, there was no damage or overheating of the wheel and the bogies. On the second test, Fulton turned the aircraft around and taxied in the other direction, reaching a speed of 122 knots. This brought the nose wheel just off the ground at about 100 knots. Fulton used reverse thrust and slowed down putting on the brakes at 20 knots and stopping. The third test was supposed to imitate an aborted takeoff. The run used the entire 15,000 ft runway. Fulton brought the aircraft up to 130 knots and cut the engines to idle, pulling back on the yoke, Fulton used the elevons and brought the nose of the aircraft up 5 degrees. The 747 acted perfectly. The nose wheel remained off the ground for some 15 feet before Fulton put it back on the runway. He then put on the reverse thrust and applied the brakes at some 40-50 knots. This really put the test to the brakes and the wheel assemblies. These maneuvers showed that the 747 held her own throughout the tests and was the perfect vehicle that would make sure the Orbiter would be safe.

Airborne

On February 18, 1977, the ultimate test was made and the pair, Orbiter Enterprise and the 747 took to the air together for the first time. It was an eventful sight and one of real beauty. The first of the flights were ultimately successful along with the rest of the test flights. Considering the takeoff weight for the pair was some 625,500 pounds with Enterprise weighing some 143,600 pounds. Fitz Fulton, having had success with the XB-70 was no stranger when it came to flying big aircraft, the SCA was the ultimate. Running many of the tests from short takeoff to short landing, Running on one engine short and many other test techniques Fulton felt that the aircraft with the Orbiter in tow was a success. However, there was no word on the Enterprise, which was just ballast at this point. There was no crew and nothing happening with the Orbiter except to hitch a ride. The testing of the Enterprise was soon to come.

Enterprise comes alive

After three months of taxi tests and flying with the Orbiter, it was time to see if Enterprise could carry her own weight in a glide test. However, there were some issues that had to fixed before that could happen, It seemed that elevon actuators had a problem and needed to be returned to the manufacturer. There were also some leaky seals and in the meantime, astronauts Engle. Haise, Truly and Fullerton continued their practice sessions in the Gulf Stream aircraft that had been fitted out like the Enterprise.

Deke Slayton, who was the head of the astronaut program, was working to set a confirmed date of June 17, however more problems arose that had to do with the primary flight control and ejections seats. These issues were fixed shortly and the next day, Haise and Fullerton finally stepped inside the Enterprise while she sat on the SCA.

The first flight had the SCA and Enterprise airborne for less than an hour. It was just to get the feel of the SCA and the Orbiter together. Johnson Space Flight Center had control of the first of the shuttle missions, the very first. It checked out in communications and so far everything was good. Ten days later, the second captive flight took place. Manned by astronauts Truly and Engle, flutter tests were made. The next test would be the big one, separation of the orbiter and the SCA. Fitz Fulton at the controls of the SCA made the shallow dive, cut the engines to idle, and opened the spoilers. In the Enterprise, Truly and Engle had been busy setting their controls for release. This dry run was the set up before the actual flight. The third flight again, a rehearsal for the big separation went the whole hog in testing everything. Haise and Fullerton were running this flight. Fullerton found that an APU had indeed overheated. However, it turned out that the problem was a sensor not the unit itself. They continued the flight. This was the end of the dry run tests. On August 12, 1977, the SCA rolled to the east end of the runway 22 at

Edwards AFB, Dryden Research Center. The SCA pilot, Fitzhugh Fulton, called for full power to the engines and took to the runway, lifting off with Enterprise on the SCA back.

Two T-38 chase planes followed the takeoff. About 48 minutes into the flight, the SCA and Enterprise were just east of the Rogers Lake. Fulton brought the 747 into a shallow dive. Fred Haise, on board the Enterprise called

"The Enterprise is set, thanks for the lift". He then hit the separation button which blew off seven explosive bolts and the Enterprise was fully and finally airborne and on her own. The 747 got out of the way by rolling in to a diving left turn. Haise brought the Enterprise pitched up to the right. Enterprise was riding comfortably. She would only remain aloft for some five minutes and start her sinking glide back to earth. Enterprise flew over Leuhman ridge, crossed highway 58 and turned to the west and Peerless Valley. She then turned towards the north side of Edwards AFB and started her line up to land on runway 17. Back at Houston mission control, they told Haise that was in a lower lift to drag ratio that actually turned out to be wrong. It was precisely on the money, so that was one on Houston.

Because of that error, Haise arrived with the Enterprise on the runway approach both high and fast. She would need the entire length of runway. Haise set the speed brakes and started his landing flare and Enterprise landed successfully after being in the air for some 5 and a half minutes all on her own. This was the true beginning of the Shuttle saga. There would be many more test flights in which the problems were ironed out. Things like control problems, short rollouts after landing due to miscalculations on descent. The fifth and last flight for Enterprise brought her to the Marshall Space Flight Center in Alabama for a series of ground vibration tests. March 10, 1978, Enterprise left Dryden for the very last time on top of the 747. Having to stay over at Ellington AFB in Huntsville over

7,000 NASA and Marshall employees saw her come in. While Enterprise was done flying, she opened the door for the real Shuttles to begin their flights to orbit.

The Birth of the Space Shuttle –Columbia's first flight

STS-102, Columbia was on Pad A, Launch complex 39 waiting for her launch into low earth orbit for the first time. The date was April 10, 1981. During some of the built in holds in the countdown, a timing fault occurred in the GPC (General Purpose Computer) While there were redundancies in abundance, this was something no one wanted to fool with. The launch was scrubbed and reset for April 12, 1981.

In the morning of April 12, 1981 under the Floridian sun and humidity, Columbia would have her next chance to show what everyone what she could do. It was also the 20th anniversary since Yuri Gagarin made his first flight into space. What an amazing technological feat when you look at the shuttle and the fact she was the most magnificent, complex machine ever created. As the countdown dribbled down to T-31, the Orbiter's internal sequencer brought Columbia to life and allowed her to take over the show. This is the system that halted the launch on April 10, 1981. Today, she was officially alive and in control, thinking on her own. At T-6.6 seconds those so deftly worked on SSMEs started to fire and ran to full power, everyone at KSC and JSC were watching their boards for any little anomaly, but there was none. The shuttle was still bolted to the pad. Astronauts Young and Crippen, on the this first "all up flight test", were the first to ever do a test flight in the U.S. It was precarious to say the very least. Crippen and Young were starting to feel the vibrations of the shuttle as she was gearing up to find her place in space. They also felt the pitch motion that cause the stack to push forward. It is a phenomenon known as the "Twang". It lasted for a few seconds as the whole shuttle stack went back to a vertical stance. At T-0, the SRBs got the order to light and there was no going back now, as the massive explosive bolts blew off and released Columbia to fly. Unlike her predecessor, the colossal Saturn V rocket who moved stately and slowly on the way to the moon, the Shuttle cleared the tower in seconds. The astronauts Robert Crippen and John Young were totally awed by their launch. Crippen's statement "What a ride!!" just about summed it up.

As soon as Columbia was clear of the tower this mesmerizing sight became even more stunning as she rolled over on her back to put her at a proper position while she rose majestically and headed up and out over the Atlantic ocean. The SRBs did their job in approximately two minutes and jettisoned off the stack to fall back into the Atlantic ocean for rescue and recycle by the NASA Liberty Star and Freedom Star barges, just waiting for them to come down.

The shuttle continued to ride away from earth on the external tank. Eight and half minutes after launch, the SSMEs were shut down after NASA "speak" MECO (main engine cut off) and the External tank jettisoned shortly after that. The ET dropped off and headed for the Pacific Ocean, burning up on re-entry.

Crippen and Young were apprehensive about the launch, feeling that they had maybe a better than half a chance it would all go alright and they would make low earth orbit without some nightmare. That was on not only the astronaut's minds, but every NASA employee, contractor, sub-contractor, right down to the janitors that helped keep everything clean, an important job at NASA. There were a couple of problems that did scare a few people. When the shuttle payload bay doors were opened to expose the radiators carried inside, it was found that some of the tiles had gotten loose from the OMS pods (Orbiter Maneuvering System) on each side of the shuttle vertical tail housing. The launch pad had been flooded with water for shock sound suppression on launch and some of the tiles got knocked off.

Critical areas were the leading edges on the wings, nose and of course the belly. There was no way to determine the damage to any tiles missing. Losing any of the heat suppressing tiles could be a major issue. It could cause a zipper effect allowing the tiles fall off and to expose the aluminum innards to super heated plasma on reentry. Essentially this is what killed Columbia in 2003. Columbia did turn her belly towards an imaging reconnaissance satellite to be assessed, but NASA has yet to release that image after all these years.

After two days in orbit, April 14, Columbia re-entered the earth's atmosphere. As she entered at Mach 25, both astronauts saw the blackness of space turn to an iridescent pink as the ionized shockwave knocked out communication for 11 minutes, the longest wait in the world. In her first flight, she gladly received the JSC Mission control call after 11 minutes of blackout. NASA and the world rejoiced but in 2003, sadly, that 11 minute blackout call from Mission control was never answered.

**The Painful Loss of Two Shuttle Crews and Orbiters. Challenger —
January 28, 1986, — What really happened?**

It was a very cold, bright morning at the Kennedy Space Center in
January of 1986. Florida is not known for icy, cold mornings, but it did
happen every once in a while. However, this cold day would prove to be
a disaster in the making. To look at pad 39B, which would be the first
time that this pad was used in a shuttle flight, it looked like a giant ice
castle fantasy that would turn into a nightmare. The night before the
launch, fights were raging between contractors and NASA as to whether
the launch should scrub or wait until later on in the day to launch Challenger and the first teacher in space. Rockwell, the Orbiter's builder,
Thiokol, the SRB contractor and Kennedy flight management were at it
big time, the issue was ice and cold temperatures. The decisions made
were wrong and we know that in hindsight. However, it didn't start right
at this point. The problem started way before the night of January 27, and
the morning of January 28, 1986. It started early in the development of
the shuttle program and in the development of the SRB. This wasn't the
first launch that was plagued by ice and cold.

STS-51C had the same issues. However, that flight was scrubbed for
a later date. Why wasn't the same procedure followed on the morning of
January 28? There is also the issue that at that time, NASA was looking
for a second source to build the SRB. This of course, was making Thiokol very antsy about their product and their part in the Shuttle program.
Thiokol was depending on the SRB as a sole source of income for the
company. Yes, it did have other projects running, but not as lucrative as
the SRB.

Second Source

Thiokol was the one and only for NASA as the supplier of the SRB. It was well known in the Shuttle program office what combined both program management functioning with engineering for each section of the SRB. The SRB manager had two duties; the first would be as a program manager and the second as a project engineer. In essence, Thiokol managers wore two hats when it came to management of the SRB. NASA initially brought on this type of management system and that is how NASA wanted it. It was cost efficient, but did it solve the issue of having a set of managers on the main engineering section that was responsible for asking the technical questions and worrying about the resolutions to that problem, instead of trying to cover two bases.

Thiokol's Space Shuttle SRM Program managers were found to be too busy trying to get the product out of the house, instead of answering the many questions asked by NASA. The component managers were below the director of the Space Shuttle SRM Program office, yet there was no line of responsibility for engineering.

The chain of command went something like this: the director of the Space Shuttle SRM project reported to the vice president of space booster programs, who reported to the assistant general manager and vice president of program management, who reported to senior vice president and general manager of Wasatch division of Thiokol. This list of corporate bureaucracy is enough to give anyone a migraine and we haven't started on NASA yet! It was amazing to note that the SRB's contract had never been held open to rebid.

Since Rockwell International, along with Martin Marietta (ET builder), United Space Booster and Rocketdyne (Another division of Rockwell that built the engines for the shuttle) all had their bid hats in the ring as far as the shuttle-processing contract went.

It worked something like this:

- Rockwell built the Orbiter
- Martin Marietta built the External tank
- United Space Booster built the components and handled the integration of the SRB, the aluminum skirts forward and aft of the external tank, the attach ring for attaching the SRB and the ET to the Mobile Launch Platform, electronics for the SRB separation and recovery system, booster separation motors to separate the SRB from the ET after burnout.

Thiokol provided some 500 people to the Kennedy Space Center for assembly of all the SRB components and to stack the SRB segments on the Mobile Launch Platform in the Vehicle Assembly Building (VAB). Thiokol was slow in trying to justify why a second source wasn't needed and their presentation was poor at its best.

Early problems in shuttle flights on the SRB

Thiokol was also finding out some things about their product that they didn't know. On shuttle flight, STS-8 (August 3, 1983) the first night launch for the shuttle program, a big problem was found in the carbon-phenolic rings that were located in the forward section of the left-hand nozzle. The rings were so badly eroded that if a motor burned another 8 to 9 seconds past the planned two minutes of burn time; the nozzle would have burned right through which would have been a disaster. This incident caused damage to Thiokol's reputation with NASA so much so that NASA decided it was time to look for a second source. Now we know why NASA was starting to shop around, there were issues with Thiokol cropping up and NASA wasn't too pleased with what they were witnessing.

A new design?

Because of the problem, Thiokol removed some of the nozzles and repaired them at Kennedy Space Center to support the next two missions after STS-8. Since it was found to be a problem with the materials inside of the nozzles, it had to be checked out and changed out which really looked bad for Thiokol. Thiokol scrambled and started looking for advice in the industry on how to best bid a vehicle to NASA by going to James Beggs, who was a former NASA administrator and now retired from General Dynamics. (Beggs was a former top management at General Dynamics before becoming the NASA administrator because the company was providing the Centaur rocket upgrade as well as a payload carrier for some potential CELV (complimentary expendable launch vehicle) configurations. Thiokol discussed the SRBX ideas with General Dynamics, which was interested in bidding concepts. Both the companies worked on concepts for a few weeks, but the U.S. Air Force said it would release the official request for proposal. General Dynamics decided not to support the SRBX.

The SRBX had three strikes against it from the onset. The first was NASA deserved the vehicle rather than the USAF. Second, Thiokol did not provide assured access to space if access was lost because of SRB failure on the shuttle. Third, NASA's pricing policy for the SRBX was not competitive. The CELV, as NASA saw it, was a way to reduce the Shuttle cost bypassing it to the USAF. Though most felt if the shuttle failed, it would be a result of the SSME (space shuttle main engines) orbiter failure and couldn't discount SRB failure. With Thiokol support, NASA's Marshall Spaceflight Center submitted the SRBX proposal to USAF, but it got no consideration. The USAF picked an upgraded Titan rocket for the CELV mission, which was later called the Titan IVA with a new seven segment steel casing version of the existing SRB (provided by United Technologies—Chemical section of San Jose California). The upgraded Titan IVA went to the Titan IVB with a three-segment graphite epoxy filament wound case version of the rocket. Further, on in this sad story, the SRM upgrade had severe development issues at Hercules Inc. and had to write off several hundred million dollars due to test failure handling including manufacturing errors and propellant mixer fires. Congress bailed Hercules out with the help of Senator Jake Garn (soon to ride on the shuttle) before his retirement from Congress.

Fire in the pit

On March 2, 1984 at Marshall Spaceflight Center. Thiokol announced a major fire has just occurred in the propellant casting area where the space shuttle solid rocket motors are manufactured back in the Utah plant. The early morning radio reports had several explosions and a large number of dead and wounded. This was according to Brad Parker, manager of the Thiokol field office at Marshall Spaceflight Center.

It was found out later, that no one was killed, and only minor injuries occurred, thankfully. However, the propellant-casting house was totally destroyed. Two shuttle segments had been lost, two casting pits also were trashed, and the new casting pit for the shuttle segments was severely damaged. The fire started in one of the four casting pits located outside, not in one of the four pits inside the brand new facility. The damage from the fire was huge and the heat intense as steel beams had literally melted. The place where the fire started was a mobile casting building that was rolled into place on rails, directly over the casting pit during the propellant operations. One of the shuttle's segments was completely cast when the fire started, so this was a huge financial loss. The propellant cast into steel segments from the propellant mix bowls that each container held approximately seven thousand pounds of propellant. It took some forty of these propellant mixes to fill one casting mix.

How the SRB is assembled

Each SRB was assembled from four center segments, a forward segment containing igniter, two identical center segments and aft segment containing nozzle assembly. Because some of the propellant dripped onto a rail by the casting pit, the propellant ignited and that was it. The fire alarm was hit and nearly 1/4 million pounds of uncured, burning propellant was lit off from the segment. Three hundred tons of propellant was lost. Thiokol was out of production for three months before a sufficient number of changes were made to the hopper transfer process of the other casting rows, which allowed safe casting of the SRM until facilities were remodified.

It was at this point that talk of the second source for an SRM was getting louder at NASA. It was obvious that NASA was not happy with all that occurred at Thiokol and the fact it was putting the shuttle program behind again. The fire and the lack of understanding of the STS-8 nozzle erosion was something NASA could not ignore. Thiokol's initial assessment that the STS-8 problem had to do with the carbon-phenolic nozzle parts was not good enough for NASA.

In Thiokol's view, the nozzle's carbon fibers break leading to "pocketing" or "spalling" (chunking out) of the rocket's nozzle hot front surface. Early static test before the first shuttle flight in 1981 showed similar anomalies in erosion that was noted in the same location in the nozzle. Thiokol engineer's thought that the problems were due to the way Thiokol laid up the carbon-cloth phenolic tape that was used to fabricate the nozzle parts instead of a material problem, which is what NASA, claimed was the problem. Both NASA and Thiokol agreed that a new design with a change in the angle of how the tape was wound on the nozzle would be the solution. A graphite-epoxy filament wound case called the DM-6 was used in a demonstration in fall of 1984.

NASA insisted that the Thiokol field office at Kennedy Space Center remove the nozzle immediately after the boosters were returned to Hangar AF at Cape Canaveral after launch. The space shuttle solid rocket motors were manufactured in four segments and transported to the Cape via railroad cars from the loading facility in Corrine, Utah, near the Thiokol plant. The segments were stacked vertically on the mobile launch platform in the VAB at Kennedy. They were joined together with the infamous "tang and clevis joints" in grooves that sat two fluorocarbon O-rings designed to seal joints between the propellant sections. Hence, the joint isn't made in the factory in Utah but actually done right there at Kennedy Space Center. It was known as a "field joint". The joints were pinned together with one hundred and seventy-seven high strength nickel alloy pins with steel bands fastened around the pins. The space shuttle man-rated system had two O-rings were provided for a redundancy in sealing because of the level of criticality it was given in the NASA list. Of course, as all have found out, a gas leak in this joint would be cataclysmic.

A Statement from NASA's internal side

Before we get into the issue of what really started the ball rolling with the SRB and ring. J.R. Thompson[14] was responsible for the design and development, test and operation of the Space Shuttle's Main Engine. Thompson brought to the forefront what the issues were while the shuttle was in the design process. Truer words could not have ever been spoken. He knew about the inner workings and problems of NASA from the inside out. Thompson explained the many problems that NASA and Rocketdyne had with the SSME when they started out working on it. There were 14 engine explosions, problem after problem with bearings

---

[14] J. R Thompson – NASA SSME engineer design and development test and operation.

burning out and high vibration issues. Engines had to be returned to Rocketdyne to be rebuilt and fixes made to solve the many problems. It took 9 months and Rocketdyne had even brought in some outside experts to help. There were over 1,000 little gremlins that needed to be addressed before the SSMEs were worth their salt. The turbine blades had fatigue cracks, there were splits, ball bearings were burning out after 2.5 seconds of run time. It didn't take long for very high vibration to show its face and shut the whole test down. We are talking seconds not minutes. Because of the fact the SSMEs were a huge part of the program, they were also the item that got the most attention. According to Thompson, because of the fact that they were in the middle of a development program, there were budgets to contend with. Dates for the shuttle's first flight was changing every other day, it was hard to keep up with and on top of that. Engines had to be rebuilt at the Michoud plant in Louisiana, while Mississippi was doing more testing. All is all, every time one of these little "gremlins" made an appearance on the engines, it was like a horse kicking you in the backside. You would have to start all over again and again and again. It was a stomach twisting nightmare. However, John Yardley, who was the Deputy Administrator for NASA at the time, had a different outlook on the subject. Yardley imparted to Thompson and other engineers to keep testing till the engine, or the turbine blade in this case, failed. With this mindset, at least you had thrown everything you had at the problem, caused it to fail and knew exactly when and where it would. That would allow you to walk back from that point and certify the engine, knowing when and where it was going to fail. A perfect example is, HAL the computer from the movie, *"2001 A Space Odyssey"*[TM15] who said that the AE-35 antenna would fail. Okay, maybe it was rigged to fail by HAL, but the idea was to put it back until it did

---

[15] Metro-Goldwyn Mayer Pictures-1968

fail, and then find out what to do about it. That is basically what Yardley was thinking. Let the engine go down after pushing it to and past the limits and then let everyone KNOW what those limits were. This way you could know what the last word was before you got into trouble.

Because the SSMEs were so very urgent on the list of things to do, they got the money and wherewithal to certify the engines, knowing everything the SSMEs would or would not do. Not so in the case for the foam strikes or the problem with the O rings. In both these cases there were signals something was not right. There were many observations and cases noted that something was wrong, neither NASA or the contractor did anything much about it. Thiokol and NASA never tested the O rings far enough to failure to see what exactly would happen if they, in fact, went over the edge. They waited till the morning of January 28, 1986 to find out. It was much the same thing with the foam strikes and how it hit Columbia on the morning of January 11, 2003. The foam breaking off was no secret. It had been happening since the first of the shuttle flights. Again, no one bothered to really explore what would happen if a piece came down while the shuttle was taking off, and as the fates would have it, the Orbiter Columbia ran right into the foam as she lifted off the pad, on the morning of January 11, 2003 and died on February 1, 2003 on re-entry. The foam breakoffs which were happening almost every launch since the shuttle's first flight were really not explored, it was taken for granted, much like the O rings. Why? That is the optimum question. It seems that in different sections of the shuttle program, neither NASA or the contractor went the extra mile to find out what the breaking point was. NASA and the contractor didn't do it, and while they hoped for the best in these two cases, the worst that could happen, did happen. This look into the minds of two of NASA's best engineers and administrators should help to clarify what really went wrong.

When the problems really started

In the SRB, system the first of the O-rings to feel the heat so to speak was designated the "primary O-ring". The secondary ring was the redundancy that in the case of Challenger did not work. Erosion of the primary O-rings in one of the SRB joints was seen in a flight close to STS-51L. That flight was STS-41B, which was launched in February 1984. In the NASA hierarchy of Levels, it was addressed in the Flight Readiness Review, known as the FRR. As a critical anomaly seen in the early shuttle program, as early as the second flight some two and a half years earlier in November of 1981, both Thiokol and NASA were not pleased with the FRR process in catching this issue. FRR's also dealt with any previous problems or observations on any anomaly that reared its ugly head. This meant anything that showed up in an inspection or returned hardware from a previous flight or any other suspect condition noted from a contractor or from issues occurring in a former FRR.

Because of the serious nature of the nozzle erosion problem, NASA requested that Thiokol look into the subject on every FRR relative to the nozzle materiel used and any inspection results of a prior flight.

O-ring erosion needed to be addressed if the erosion was noted in prior flights. The FRR process was mind boggling to say the least. The FRR was a series of eight independent reviews before a shuttle flight they listed this way:

An internal review with Thiokol;

• A review at NASA Marshall Space flight Center with the SRM program manager,

• A board meeting at Marshall with the SRB Program Managers Board,

• A meeting at Marshall with the shuttle project managers review board

• A meeting at Kennedy Space Center with NASA

• The Associate Administrator for Space Flight Review Board consisting of NASA managers and directors from Marshall Space Flight Center, Johnson Space Center, Kennedy Space Center and NASA headquarters. A pre-launch review with the same board two days before scheduled launch.

• A final pre-launch weather briefing the day of the launch.

Just looking at this list would give anyone a headache, yet this was how NASA got its work done. The entire process started six weeks before launch. Any open issues were addressed at the LEVEL 1 Flight Readiness Review (FRR), the day before launch. The FRR took a tremendous amount of time and energy for all the top NASA managers to get into line and contact other managers to finish out their workloads. The Review was a stand up, oral presentation complete with viewgraphs and charts, along with printed bound copies of all the materiel distributed to the meeting attendees. There were dozens of people attending these meetings along with the Review board. When you consider that one internal review would take at least ten hours is overwhelming. The discussion about STS-8's SRB pocketing erosion problem was on the list. The O-ring erosion however, was not getting the same attention, the nozzle erosion problem was. Through 1984, the nozzle erosion, not the O-ring was the most noteworthy problem.

A new filament wound case (FWC) for the SRB, was developed by NASA for the U.S. Air Force to help with the Vandenberg facility in California.

The substitution of graphite fibers for the cylindrical steel section of the SRM motor case dropped the weight by 28,000 lbs. allowing a 5,000 lbs payload bonus. Hercules Inc. who was manufacturing the casing, sent the first segments off to NASA.

Tongue twister.

Anyone who has ever seen a technical report from NASA would get a migraine trying to figure out the intense jargon that precedes any recognizable word. It is worse than what happened with the Tower of Babel. You can't make a full sentence at NASA without using acronyms like: **STS, VAB, MLP, SRM, SRB, SSME, ET, MECO, FRF, FR, HPU, etc**.

The best that this author has heard so far is, are you ready? SQUATCHLOIDS! Before you run out of here screaming, it means the envelope of loads occurring on the shuttle vehicle during the ascent flight. This envelope includes flight trajectory, seasonal wind loads and three SIGMA dispersions in the SRB thrust mismatch, thrust variations, aerodynamics and flight control system variations.

If everything predicted in flight loads was within this envelope SQUATCHLOIDS were "Green". In short...go figure. Is it at all possible that this tendency to use so many acronyms will have caused undue stress in attempting to dissect a report? Remember, these reports were read by contractors and others that may not have been privy to the entire acronym library that NASA had devised. When you can't understand what you are reading, it can make your job more difficult, tiring and to say the least, confusing.

NASA Review Boards

In keeping with NASA protocol, NASA did form the *Senior Material Review Board,* better known as the SMRB at the NASA Marshall Space Flight Center and at Thiokol. This board approved or rejected all discrepancies in critical hardware plus what's outside the SRM's base. O-ring erosion was not happening on every flight and it wasn't considered by Thiokol or NASA to be a critical problem, at least not like the nozzle spalling which was still the big issue at the Flight Readiness Review (FRR). O-rings were on the back burner. It was still an issue, but as the O-ring problem arose in conjunction with the nozzle issue, Thiokol's first (filament wound case-FWC) FWC-SRM Development motor, the DM-6 was being prepared for static testing in the early fall of 1984.

The FWC had two design features, which might improve the "field joint" O-ring performance and help the nozzle erosion problem. FWC field joint included a metal capture on the tang (tang and clevis joint) side of the joint that reduced the rotation or opening of the joint during pressurization This would make it an issue for the O-ring to hold a seal and was part of the Hercules company design for the FWC. The FWC motor had design changes in nozzle to eliminate the pocketing erosion or the carbon-phenolic inlet rings of the nozzle. The Aerospace Safety Advisory Panel (ASAP) visited Thiokol in June of 1984. They wanted presentation on design of FWC/SRM Tests program and what would lead to the certification of the new FWC-SRM motor for flight. ASAP wanted to review the FWC-SRM and no discussion of the O-ring pocketing erosion noted on the nozzle of SRM's flying on the shuttle.

The DM-6 motor was tested October 25, 1984. It was a full ground test of a full-scale space shuttle SRM, new graphite case version. NASA determined it wasn't necessary to conduct any more ground tests after the motor was flying on the Shuttle unless there were major redesigns of one or more components, which was now the FWC (filament wound case). The major competitor Hercules Inc. was fabricating the large graphite cylinders under a sub contract with Thiokol. The company sent a group of people up from the other end of Utah to see the DM-6 inspection. It looked good; the nozzle was going to make it. The test results showed that the nozzle was in good condition from anything previously tested, with no spalling.

NASA was pleased to say the least. This FWC-SRM contract was separate contract with Thiokol and was a cost plus award contract, which allowed NASA to evaluate the performance against the material planned for that time span. NASA awarded the FWC-SRM a top rating for the evaluation. Thiokol was proud to say the least. However, Hercules was not so happy with the outcome from Thiokol due to contamination problems, late deliveries and cost overruns. It was so far, so good for the FWC, since there was no holes or voids that were blown through the putty. Thiokol was sure that during the mating operation, air would get trapped between the seals, but pressure could blow holes through it. It turned out that using a broomstick and putty to fill in the voids solved the "problem" of holes. It also turned out that every test field joint was treated the same way. This "procedure" did not show up in any manuals procedure reports.

Problems

Disturbing data from FWC-SR test came when he igniter was removed from the steel forward dome. After the segments were removed from the test bay, Thiokol saw the primary gasket seal eroded totally at a location that was in line with a blow hole in the vacuum putty that was used to protect the seals in all the major joints. There was black soot between the burned primary gasket seal and secondary seal. This was serious news. Thiokol dissembled the rest of the nozzles and igniters and found soot behind the primary gasket seals in four of the last eight flights flown: STS 41-C, STS-41D, STS –51 D, STS-51 G.

In each case, soot was found where there was a blowhole of some type and that was repaired with a broomstick and vacuum putty through the joint. None of these flight motors had shown any erosion in the primary seal as shown on the DM-6.

Thiokol checked all manufactured assembly records and saw no outstanding issues. The seal was certified before shipment to Thiokol. With further inspection, it seemed that the seals were out of calibration. They had more rubber in the sealing area than the specifications permitted. This extra material on the gasket diminished the preload on the attachment bolts during the fast pressurization building up in the igniter and that caused a deflection between the adapter and metal gasket retainer. This allowed the blow by to miss the primary seal during ignition. All the secondary seals looked to be all right. Thiokol dumped all the seals that were miscalculated and made during the time the inspection took place.

The Flight Readiness Review accepted this so that future shuttle launches would use the "used" gasket seals for a while. This was an acceptable situation for NASA and Thiokol since the SRM was really they only "reusable section" of the SRM. This test on the FWC (filament wound case) made everyone happy so far. It should be noted that the man who found and figured out what was wrong was Alan McDonald of Thiokol.

Some real serious problems

Life wasn't all a bowl of cherries, however, when it came to the DM-6 and the FWC- SRM. The nozzle was removed from the aft section that arrived back at Kennedy Space Center after recovery of the STS-41D flight with Discovery, August 1984. There was erosion of the primary O-ring of the nozzle to the case joint, seen on one of the two booster nozzles. The amount of soot between the nozzles primary and secondary O-ring was menacing. It was the first time this was seen in the area of any O-ring seal within the solid rocket motor.

Seeing soot behind the main O-ring was scarier than the erosion of the O-ring itself. It showed the O-ring primary had failed its main job to prevent hot gases from passing through the seal. The secondary O-ring held its place. This was a frightening discovery. The next shuttle flight coming up STS-41G Challenger was due on Oct 1984.

To be ready for this flight, Thiokol started to figure out explanations for the soot and how it could lead to a catastrophic failure. Thiokol also had to have reasoning for possible delay of the flight until the problem could be corrected. There wasn't a lot of time to do it in. The examination of what the Discovery flight brought back showed that erosion of the primary O-ring seal was "self limiting" according to Thiokol.

The hot gas erosion caused by the blowhole in the vacuum putty and joint, during the ignition and encroaching on the O-ring in a localized area of the just down from one of those blowholes. This process was self-limiting because the action stopped as soon as the motor got to a steady pressure and stopped the differential of pressure during the ignition process. The whole process took 0.6 seconds after the ignition. Testing showed that half of the O-ring had to be eroded away before it failed to function as a pressure seal. The erosion of the O-ring on the STS-51D flight was 0.058 inches or 20% of the cross section, which was 0.275 inches in diameter.

Thiokol felt there was enough protection for the O-ring so long as the O-ring extruded into the gap between the metal parts in the joint before any real amount of hot gas could get through the O-ring. The real headache was that the soot blowing past the O-ring during joint sealing process. What was confusing was the soot between the primary and secondary O-rings wasn't lined up with the blowhole through the vacuum putty in front of the O-ring that was eroded.

Even after the joint was made, it was tested with nitrogen pumped through it. That would cause the O-ring to move into position and the secondary ring to move back to seal when pressurized by hot gas. Thiokol felt the O-ring did not move back into the groove evenly because at pressurized differential around the 300 inch circumference of the joint and small amounts of soot and gas must have blown by the O-ring during the movement process.

Thiokol also felt the amount of gas wasn't enough to damage the O-ring before it was fully sealed and seated. No soot ever reached the secondary ring. Thiokol felt after all this inspection and testing, it was still safe to fly the shuttle.

Of course, during a briefing with NASA there was "bone of contention". The Director of science and engineering at Marshall Space Flight center challenged the Thiokol theory straight out and said they did not believe them. To say the least, this upset Thiokol but NASA wasn't budging. They walked out of the meeting, not recommending the next shuttle flight with Challenger.

Larry Mulloy, who was Marshall's project manager for the SRB felt that the Thiokol presentation, which many walked out on, should continue. It seemed that the director of science and engineering for Marshall Spaceflight Center was bit miffed by Thiokol's request that if they didn't believe them, maybe they would take over the presentation. This put the NASA managers on the spot. However, being in a powerful position in NASA, that just wasn't done. Mulloy recommended that the same presentation on the issue be given to a LEVEL III Flight Readiness Review the next day. In that next day's meeting, the FRR board did agree with Thiokol conclusion. It just showed how easily some one's nose would be severely unseated if they felt that they were being challenged in any way. This was just part of the NASA political and egotist process.

Where the problems *really* started and how STS-51L was lost. Signs that were not seen and what the astronauts never knew.

Shuttle Flights

There were four shuttle flights in 1984:

STS-51A Discovery: The O-rings showed no erosion or soot blow by in any field joint. The Nozzle showed minor pocketing erosion.

STS-41B Challenger: The Primary O-ring in left forward field joint had eroded 0.040 inches and primary O-ring in the nozzle to case joint on the right hand booster exhibited 0.039" of erosion.

STS-41C Challenger: No erosion on primary O-rings in aft field joint of left hand booster. There was a blowhole through the vacuum putty showing heat effect. 0.034 inches of erosion was shown on the primary O-ring of the right hand SRB nozzle and soot blow by happened in the igniter to case joint of the same booster.

STS-41D Discovery: The primary O-ring of the left hand booster nozzle joint showed 0.046 inches of erosion while right hand booster showed 0.028 inches of erosion on primary O-ring in the forward field joint along with soot blowby on the same booster past the primary igniter gasket seal.

In: STS-41B—Ron McNair

STS-41B-- Dick Scobee

STS-41D—Judy Resnick

STS-41C—Dick Scobee

These astronauts flew in prior flights before being joined to the Challenger crew STS-51 L. The shame of this is the astronauts NEVER KNEW the situation of the O-rings and the nozzle joint. They were never told the seriousness of the problems facing the SRB. Later on, Rocketdyne, who manufactured the SSME were always finding turbine blades in the engine damaged. There was turbine pump damage from flight operations to test stand, failures. The problem was bad enough that the Shuttles main engines had to be changed out. Again, the astronauts never knew the seriousness of the problems that the shuttle and the program were facing. While the astronauts were all astute engineers and scientist, they relied on NASA to take care of the shuttle and her issues.

After the STS-51C mission, Discovery had the SRBs returned to hangar AF at Cape Canaveral there were more serious problems to deal with. Two of the field joints, one on the left hand side and one on the right had side, showed erosion of the primary O-ring and considerable black soot. Remember that STS-51C was launched on a cold day in January of 1985. The soot showed up between the primary and secondary O-rings in both the field joints. Here was the first "smoking gun" if you would forgive the very unfunny pun. Both of the nozzles showed soot between the primary and secondary O-rings without any erosion of the primary O-ring. While the O-ring held the seal after the initial blow by, there was no serious damage to the secondary seal. However, the secondary seal did show heat damage. As mentioned, this was the coldest day to date for a launch in the shuttle program. Even though launch time temperature was 62 degrees, the O-ring temperature was calculated at about 53 degrees. She had been sitting outside three days before launch and it was the coldest three days on record for the Florida coast, with temperatures in the teens during the nights. The launch of STS-51C had been held a day from January 23 to the 24th because of the cold weather. This begs the question what happened to 51L. NASA was growing concerned about the potential for extreme icing conditions on the external tank and the launch pad made matters worse because of the various water suppression systems used during the launch. While the weather did warm up during the launch day, the extreme cold weather at night had kept all the hardware on the pad at below comfortable temperatures, which made it tougher for the O rings to seal properly.

Again, if this information was assessed and digested before the 51L flight, why wasn't this information used in processing the yeas and nays for launch on January 28, 1986? At the next Flight Readiness Review (FRR) meeting, Thiokol tried to explain what they saw on the 51C flight as serious problem. The exact statement: "Launch of STS 51C was preceded by the coldest three days in Florida history" They recommended the next launch STS-51 E which was really STS-51D be set for April where temperatures in the teens wouldn't be an issue.

The program goes on

In April of 1985, STS 51D was to launch with Senator Jake Garn on board as a payload specialist. The post mortem on the SRBs showed that the primary O-rings in both the rocket nozzles. This was a first time occurrence. The erosion in the right hand nozzle was 0.068", at least a third more than ever seen before. Conditions on STS 51B showed erosion of the primary O-rings on both nozzles but also that the nozzle on the left side had experienced a serious problem.

The primary O-ring on this nozzle had failed to seal at all. It had eroded completely through in three locations with the worst location showing only 1/3 of the O-rings original cross section still remaining. There was a very heavy coating of soot between the primary and secondary O-rings yet 12% of the cross section of the secondary O-ring had eroded away.

This was pretty frightening information. However, because of the long turn around time in recovery of the boosters and hauling them back to KSC to have them torn down and checked, the serious O-ring conditions weren't found until late in June of 1985 after yet another flight had gone off (STS-51 G-Discovery). STS-51G showed marked evidence of erosion of the primary O-rings on both nozzles. The condition on STS-51B was serious enough to warrant a full investigation and assessment as to whether the condition could worsen to a point where the shuttle was not going to be safe enough to continue flying. Here are the beginnings of the serious issues regarding the primary O-rings and the secondary O-rings. Given the turnaround time to process the returning flights, was there any way that this information could have gotten into the system to protect a flight that was about to go? Was the bureaucracy of NASA so formidable that information couldn't get to the people who needed to process and evaluate it before the next launch was ready to fly? This question needed to be asked.

Conditions

When the shuttle's main engines started up, the space shuttle literally bent over, which placed some very high stress loads on the shuttle stack, especially on the aft skirt that supported the entire 4.5 million pounds of the shuttle. To prevent the shuttle from tipping over on the pad, the skirt was bolted to the mobile launch platform. The top of the SRBs were displaced by more than two feet from the vertical and high loads introduced to the SRB center and aft field joint, which put one side of the SRB in total compression and the other side in tension. It looked something like this, imagine you are standing and leaning on a cane, you bend backward and lean on the cane, the pressure on the cane tip increases for the entire cane, while the tension of your body leaning on the cane tip is stressed. It is easy to imagine how much force was transferring back and forth in the shuttle stack. May 1985, the last development motor static test of the DM-7 was going for design certification and delivery to Vandenberg AFB for the VLS-1 or the first shuttle flight out of Vandenberg. Of course, this was all an Air Force deal. The case to nozzle joint did exhibit some rather bad erosion issues in the primary O-ring. Since this nozzle was disassembled before the ones that were returned from

STS-51B, it was the worst erosion that was observed on a primary pressure seal up to then.

Larry Wear, who was the SRM program manager at Marshall Space Flight Center. Wear said that the USAF had asked NASA for a briefing on the status of the shuttle's SRB program with emphasis on what had been learned and what could be applied to the new CELV (complimentary expendable launch vehicle) program which was in the development phase. The USAF wanted to know about the O-ring problems, nozzle erosion and such. Larry Wear said the USAF asked for a tour of the shuttle manufacturing facility in Utah and a briefing on the SRM program. The visiting team from the USAF was well aware that Thiokol was in competition with United Technologies Chemical Systems, division of United Technologies. The SRM contractor for the CELV was the second source for the space shuttle solid rocket motor.

Since the Shuttle SRB were reused, it was necessary that a good coating of HD-2 (hydrocarbon grease) was used on all unpainted surfaces to prevent salt water erosion when the spent rockets dropped into ocean after burnout. The grease was more effective that the original weather seal.

All that seal managed to do was to trap seawater in the joint during splashdown and recovery, which made the corrosion worse. It was eliminated from the SRB maintenance program.

Disasters for Thiokol

The year 1985 was not good for Thiokol. The first of the O-ring erosion and seriousness of them was showing up in post flight checks. There was also the problem with the nozzle joints STS-51C showed that cold weather and O-rings did not mix. There was blow by and loss of the primary seal by erosion. There was also erosion in the secondary O-ring in STS-51 B. That wasn't a cold temperature launch.

January 3, 1985 started the year off terribly for Thiokol. A rare thunderstorm in Utah caused a mix of propellant to go up in flames with a lightning strike in the mixer building. 7,000 lbs of shuttle propellant ignited and destroyed the entire building. Fortunately, no one was killed. The NASA investigative teams arrived to check out what happened. This caused Thiokol some serious grief because Thiokol did not have any "failure modes, effects analysis, or hazard analysis" to offer. It was clear to NASA that the redundancy offered was a 600-gallon mixer that reduced the chances of the entire Shuttle program going down to a mixer fire. Meanwhile, STS-51B was giving fits to Thiokol because the primary O-ring and the left hand nozzle had never sealed and probably leaked during the leak check that had been conducted after the initial installation of the nozzle on the motor in the factory. STS-51B's O-ring erosion could only have happened if the O-ring was not capable of sealing at all and that Thiokol had not detected the leak during the check was because the vacuum putty had managed to hold the maximum leak check pressure of 100 psia that was pumped into the joint.

NASA asked Thiokol what they knew about the primary O-ring failure on the nozzle of STS-51B. Thiokol told NASA what was already known. NASA wanted a Thiokol team to come to NASA headquarters in Washington D. C. to discuss the entire event. Larry Mulloy, (SRB project manager at Marshall Spaceflight Center) found out about the request and got upset. He responded with "Headquarters people should not be contacting you directly about that."

Mulloy was angry, he said, "They should be coming directly to Marshall for that information." Here we have the proverbial peeing contest about who should know what when. Marshall Space Flight Center was very jealous of their position, as were all NASA centers. This was part of the problem of how things happened or did not happen.

Mulloy had said Thiokol should do a first class presentation for NASA headquarters but make sure that the presentation was reviewed first by Marshall Space Flight Center BEFORE going to Washington. Mulloy said, "I'll coordinate this with headquarters and set up the meeting date."

Was it necessary for a project manager to set up a presentation for headquarters? What was it that Marshall didn't want NASA HQ to know? Alternatively, was it just NASA politics in full bloom?

The O-ring erosion problems were getting progressively worse, yet the shuttle continued to pump out missions as if all was perfect. The loss of the primary O-ring with major erosion of the secondary O-ring in the nozzle joint on Challenger (STS –51B) flight on April 29, 1985, raised the level of worry so much that Thiokol decided to form an internal O-ring task force in July of 1985.

An additional review of the problems of 1985 was set for NASA Headquarters.

Shuttles still up and flying

The shuttles continued to fly:

STS-51 G June 1985

STS-51 F July 1985

STS- 51 I August 1985

STS-51-J and STS-61 A in October 1985

STS-61 B November 1985

STS-61 C Jan 1986 flew with Congressman Bill Nelson. It was scrubbed during the final seconds of countdown on December 1, 1985 but launched January 12, 1986.

From April 12, 1985 to January 12, 1986, there were nine shuttle launches. It was one launch per month. Nevertheless, not all was a rosy picture. The post flight examinations of the returned SRBs, showed that hardware from the above flights had nozzle joint and O-ring erosion. It was no longer just an occasional problem it was now becoming a chronic problem. All nine of the flights had problems, O-ring erosion in the case field joint. That is the joint that is done at Kennedy Space Center, delivered in a vertical position to the center, was eroding. Thiokol knew there was even erosion in the field joint in STS-61 C that flew in January of 1986. There was soot blow-by in two joints of STS-61A on October 1985. With this entire happening, the question needs to be asked...what was NASA doing about it, what was Thiokol doing about it? Why didn't the astronauts know that they were sitting on top of such a potential disaster?

Larry Wear (Program manager for the SRM) made the statement to Alan McDonald of Thiokol (*Truth Lies and O-rings - Alan McDonald's excellent book*) *"Al. You know there is a real chance of a shuttle failure. It's going to happen eventually. I think it's inevitable. Most likely the failure will be due to a problem with one of the high speed turbo pumps in the Space Shuttle's main engines...."* Wear had seen many catastrophic failures during the development of the SSMEs and he knew that Rocketdyne was still having problems with engine failures on the test stand. Hence, it was not only the O-rings and Nozzles that Wear was worrying about, the space shuttle main engines were also a huge cause for concern. Why was this not brought out? Wear went on to say; *"NASA is not prepared to deal with a shuttle failure and neither am I, no matter what the cause. If a disaster comes as a result of an SRM failure, now that I am program manager for the SRM, it'll be even more difficult to deal with"*. It was obvious that Wear and McDonald were both having nightmares about how and when a shuttle disaster would hit. However, Thiokol's McDonald had his faith in the SRBs than in the main engines.

How could the shuttles fly with this knowledge out there?

After the STS-51B flight, Thiokol knew that the shuttles were still going to continue on schedule. The problems of the case joints and nozzles, along with secondary seals and all the rest were not unknown to others at Thiokol. Roger Boisjoly, also of Thiokol had just written a memo to his supervisor stating that NASA officials were not well informed about the issues with the O-ring seals or with what really happened on the STS-51B flight. Boisjoly knew more about the O-rings than anyone at Thiokol or NASA did for that matter. He knew the loss of the primary seal on the nozzle joint STS-51B (April 1985) was a major problem. It was only because the secondary O-ring held, that really saved the flight and the astronauts.

NASA management in Washington D.C., Huntsville, Alabama and Houston, Texas were more concerned with the pocketing erosion that was seen on the nozzle joints than they were about the erosion on the O-rings. Larry Mulloy, (SRM Project Manager-NASA) requested that Thiokol put together a presentation of the STS-51B problems for Marshall Space Flight Center to review. Marshall agreed that it was urgent for NASA officials to know what was going on and that Thiokol conclusion about the field joints was a major concern. Marshall wanted Thiokol to add a recommendation that is was safe to fly the SRBs if inspection and leak checks were done at the time of assembly. However, Mulloy and the rest of the Marshall crowd wanted something eliminated from the draft. NASA asked that the statements said: "…data obtained on resiliency of the O-rings indicate that lower temperatures aggravated this problem." Here, at this point, is where NASA's administrator should have called this entire scenario into question. How could this be accepted?

NASA continued their request with that Thiokol could leave in the statement about Thiokol's concern on joint deflection and secondary O-ring resiliency. This request was a frightening turn of events. This doomed the shuttles to fly with a major problem. Again, the astronauts knew nothing about it. Thiokol made the changes as requested. This NASA controlled presentation was made by Thiokol at NASA Headquarters on August 19, 1985. The spotlight of the Thiokol presentation was SRM pressure seal issues.

The Conclusion

In the final statement of the presentation that Thiokol gave, they felt that the primary O-ring should not erode through, but if it looks like due to erosion or lack of sealing the secondary seal will not seal the motor.

The Recommendation

Thiokol's recommendation to relieve the issue of the erosion due to lack of a good secondary seal, in the field joint (that's the joint that is done on the premises of Kennedy Space Center) is most critical. Ways to reduce joint rotation were incorporated as soon as possible to reduce the critical position on the NASA list of urgent issues to be repaired or redesigned.

Yet, if the elimination of the joint rotation could be done, both O-rings would seal under all operation and environmental conditions, which would be independent of the O-ring resiliency. The capture feature (a metal lip on the tang side of the failed joint) would restrict the joint the joint during the pressurization at ignition. No one disagreed but no one at the presentation said, "Fix it" either. It was a stalemate, pure and simple.

Why was that? Thiokol put together a full task force in the Utah facility the day after the presentation was made at NASA.

Roger Boisjoly

The man, who was blamed and scorned as being the "whistle blower", Roger Boisjoly, passed away from cancer February 3, 2012. Boisjoly knew more about the O-rings than all of NASA and Thiokol put together. He was the point man on the O-rings, and he took it seriously. To bring us up to this point in the story, Boisjoly had written a memo in July of 1985 stating his concerns about the conditions of the O-rings. Even though the task force on the O-rings which we have already mentioned was in play, it was mired in the worries Thiokol had about losing the contract and NASA's looking for a substitute, bureaucratic tie-ups with not only paperwork but personnel and NASA's frozen attitude that cold had nothing to do with launching a shuttle. Boisjoly paid a stiff price for his ethics. After the Challenger explosion, Thiokol and NASA demonized him, because he turned over internal documents to the Rogers commission showing just what was really happening with the O-rings. His grief was so intense; Boisjoly spent the days after the disaster moving rocks around in his backyard, unable to function mentally or physically. Boisjoly was of as an unsung hero in trying to alert NASA and Thiokol to the danger that they were facing. However, it wasn't heeded.

Boisjoly's memo

Boisjoly was so concerned about the problem of the O-ring, he wrote a memo to his vice president in Thiokol, Bob Lund in July 1985. Lund was in charge of engineering. Boisjoly's memo read:

"This would result in loss of the shuttle vehicle and its crew" this letter was largely responsible for getting a formal task force assigned to the O-ring seal problem with Boisjoly heading the team. The Task force formalized at NASA in August of 1985, after the NASA presentation. In September of 1985, the task force submitted its first recommendation to NASA/Marshall for solving O-ring seal problems to incorporate the capture feature in the field joint.

After the discussion at Marshall/NASA, they formally rejected the proposed change saying Thiokol did show how the change would increase the service life of the SRB case joint, it would increase the cost of the shuttle program. NASA told Thiokol to resubmit the engineering change proposal (ECP in NASA speak) with cost benefit analysis. It was going to cost more money, period.

The NASA Engineering Change Proposal was resubmitted, yet NASA and the Marshall Space Flight Center didn't answer the resubmission until April 1986. This was three months AFTER Challenger exploded.

In December of 1985, Brian Russell a program manager working in on the ignition system and final assembly area got a phone call from Jim Thomas, who was in Larry Wear's NASA SRM Program Office at NASA/Marshall. It had to do with the Problem Assessment Report (PAS). This was a reporting system used by NASA to document every anomaly noted on a shuttle flight in post flight inspection or ground test of the shuttle propulsion component. It was a large printed document of the old 1980s IBM printout, which was so gruesome, and overwhelming, no one bothered to pay any attention to it.

More bureaucratic babble and annoyance

Further on down this tragic road, Jim Kingsbury, the director of science and engineering had complained that both the shuttle contractors and Marshall Space Flight Center were not closing out items on the PAS (Problem Assessment Reports) It was also the wish of Kingsbury that Thiokol would please reduce the anomaly list. Of course, the PAS reports contained information on the blow-by and individual putty holes in the O-ring erosion problem. Thiokol surmised it all by calling it "pressure seal anomalies." No matter what it was called, this wasn't going to change anything on the closeout sheets for the PAS. For all intents and purpose, the PAS reports were put in the circular file (aka - the trash can). In the last half of 1985, there were many new people in the shuttle program on the NASA side. The first and most important was Stan Reinhartz who replaced Bob Lindstrom as the manager of the space shuttle projects office at Marshall Space Flight Center. Reinhartz was a total neophyte when it came to the shuttle program. In his own words, he was never comfortable in the position. Reinhartz's prior post was a Marshall and he was dealing with the upper stages of rockets and satellites. He had no shuttle experience at all. To show his crew's feelings, there was a running joke: if you said anything about ET (external tank) Reinhartz ran out to "phone home". If you said LOX, he went out for bagels. Not only is this a sad statement about the morale of NASA, especially the Marshall Center, but also it proved to be a most dangerous state of affairs when you were dealing with the Space Shuttle program, something so complicated and delicate, it boggles the mind that this man was put in charge. The Shuttle Project Office Manager's position was the most important position in the Marshall house. It was obvious in all the FRR (Flight Readiness Review) meetings, Reinhartz lack of participation and focus was noticeable and obvious laughable to some of the NASA employees.

Reinhartz was also the proverbial fish out of water in the FRR Board meetings that were chaired by the Marshall Center director, Dr. William Lucas. Reinhartz was leaning on his lower echelon managers to get him through the meetings, a very scary proposition.

NASA Administration is MIA

At this time, NASA Administrator James Beggs was under investigation for charges relative to his former company General Dynamics. This also wasn't good for NASA morale and it left many holes in the NASA Management throughout the field centers and system. The charges against Beggs had to do with improper use of research and development funds in General Dynamics. It was involved with a contract on tactical weapons system that General Dynamics was in competitive development. Beggs was a key General Dynamic's executive responsible for this area at the time and was investigated for his possible criminal involvement. Beggs temporarily resigned from NASA administration and was replace by Dr. William Graham. Dr. Graham was also a shuttle neophyte. His background had to do with "black weapons" development for the Department of Defense. Graham was hardly the man to give the reins of the shuttle program and NASA to, at the drop of a hat. Beggs wasn't too pleased with Dr. Graham's appointment.

Beggs felt that Graham was not qualified to lead NASA and his background was far from an asset to NASA. However, Dr. Graham was in the game. Even though Beggs was found not guilty, his reputation was so damaged there was no way he could return to NASA's front seat.

Challenger on the pad

STS 61 A, which was the shuttle Challenger's flight, was flown in October of 1985. The post flight inspection of the SRBs showed soot between the primary and secondary O-ring in two field joints on the same motor. While the amount of soot, wasn't as bad as it was on STS-51C (which was launched on a cold January day in 1985) STS-61A showed no evidence of soot between the seal on the nozzles where soot had been seen in STS 51C. Soot between the O-rings in STS-61A field joints were a light gray rather than the thick, black soot that was shown on 51C. There was no O-ring erosion of 61A in contrast to the severe erosion on both primary O-rings of STS-51C. The reason had to do with temperature. STS 61A was launched in 75-degree weather, while STS 51C had flown at 53 degrees. Temperatures were making a difference, no way around it, yet NASA didn't want to hear about it.

An overblown balloon

As the shuttle scheduling was heating up, the stress on Thiokol's manufacturing and resources was stretching to the limit. Thiokol was at their maximum in producing the SRM segments and was just making the load required for the increasing shuttle flights. At the time, they were trying to meet twenty-four shuttle flights a year by 1988. This was a very critical time for Thiokol because NASA was making it known that they were shopping around for another supplier and they were already evaluating proposals from Thiokol competitors to establish a second source for the SRM production line. Thiokol was working much like an over blown balloon to make the schedule shifts. We all know what happens to overblown balloons, they burst.

Thiokol was trying to hold on by submitting a "BUYIII" proposal to keep in the game for NASA production of the 66 upcoming flights. That meant 66 sets of motors with option for as many and 99 sets. NASA was going to use the proposal to compare costs with those of the competitors in the second source proposal to determine their economic feasibility.

At the same time, NASA told Thiokol they had to clean up all the back PAS anomalies of which two were serious problems that occurred at Kennedy Space Center for STS-51 L, which was scheduled for January 1986.

At this stage in the story, we now see that the problems that brought Challenger and her crew down did not happen overnight. It wasn't just the temperatures. It was the entire process. The two problems that occurred at KSC:

• Aft exit cone of the nozzle on the right hand side SRB.

• Visual inspections of the aft exit cone revealed a massive unbond of the phrenolic liner from the aluminum shell. Because Thiokol didn't have a replacement exit cone available, they sent some of their engineers to

179

KSC to inspect via ultrasound the un-bonds. The amount of un-bonded area in the ultrasound showed that it, in fact, could be repaired. Thiokol concluded that the un-bonds could be fixed using injected epoxy adhesive in the un-bond area a quick fix accomplished by Thiokol engineers and NASA engineers together.

• Improper segment handling operation at KSC.

• This mishap occurred when the handling rings were being removed from one of the left hand, center segments just before stacking it. In shuttle assembly, the handling rings were connected to a segment by the same high strength pins that hold the segments together. The pins were not removed when the crane was lifting of the handling rings, causing a high load on those pins. This resulted in a slight bending of the clevis leg and some minor deformation of the pinholes.

• A structural analysis of the SRB segment indicated that the require safety margins could be met and the segment could be used and mated with the proper O-ring squeeze at the joint.

Given these three serious issues, the engineers at Thiokol felt that the fix was solid and agreed to use the segment. However, Marshall/NASA program office decided to substitute a new segment. This new segment was added by using a center segment that was already in storage at KSC. There really wasn't an issue with the bent clevis leg because it was replaced. However, Jim Kingsbury didn't agree with the recommendation to use the repaired aft exit cone on the right hand solid rocket booster for the STS-51 L mission. This was even though it was tested with ultra sound after it was repaired. The aluminum pins that were used in the cone's phenolic liner to pin it to the aluminum shell were all right to hold the line even if it became totally un-bonded from the epoxy adhesive. Kingsbury said, *"I do not agree with Thiokol's recommendation to use the right hand aft exit cone in its repaired condition."* Thiokol responded with *"As a result of all our testing, we actually know more about the*

right hand aft exit cone than the left hand exit cone, and I personally
(That being Alan McDonald of Thiokol speaking) feel more comfortable
flying it than the other which though manufactured at the same time has
never been ultrasonically inspected and has been accept for flight purely
on visual inspection prior to assembly."

McDonald was right. Wouldn't you feel safer with something that you have seen the guts of and know it passed through your hands for repair, than something you are only estimating is correctly assembled? If it was a matter of the fact that it was repaired, just think of the everyday situation in your car. If your mechanic had repaired your carburetor and he signs off that it is safe, knowing the ins and outs of your particular car, wouldn't you feel safe with it? People do it every day. In many cases, the repaired part is more efficient and safer than something taken out of the box that could have imperfections. In the case of Thiokol versus Kingsbury, the visual inspection can detect un-bonds on the edge of the lining if they are separated from the shell. The right hand, aft exit cone un-bonds were clearly visible to Thiokol but the major area of the un-bonds was deep inside of the part. The cause of the un-bond in the right hand aft exit cone was known, but since both parts were processed in the same way at approximately the same time by the same team, it is reasonable to assume that the left hand aft exit cone may have some un-bond as well. Kingsbury fought the point and told Thiokol that he would not support the use of the repaired aft exit cone for STS-51 L.

At this point, Thiokol turned to Larry Mulloy, (SRB Project Mgr. at Marshall/NASA) "At Marshall, logic and good judgment are not always enough for making a decision. Many times, logic and good judgment would roll over for superstition and fear in the interest of avoiding pain." This is what Mulloy had to say to his boss, Dr. Lucas at the next FRR meeting. It turned out that Mulloy was right. When this was presented to the LEVEL III Center Directors FRR (Flight Readiness Review), Kingsbury had his nose rubbed in it because he had to agree with no dissenting votes. While Dr. Lucas asked Kingsbury if he was all right with the decision, Kingsbury claimed no, he wasn't. Dr. Lucas asked him if his concern was enough to recommend not using the exit cone and delaying the launch, Kingsbury said he was not for delaying the launch with the repaired cone. This shows the way that things happened at NASA. Dr. William Lucas had the modus operandi of being a tough guy when it came to delaying launches. In fact, none of his people really wanted to go one on one with him on any point. He demanded, he received, and that was the way Marshall Space Flight Center was ruled. Here we have the visual image of what was happening at Marshall, Thiokol and Kennedy Space Center: Marshall was afraid of its director, Thiokol was afraid of losing its bottom line and Kennedy Space Center was afraid of losing its schedule. Many fears but the wrong kind, they should have been afraid of what was happening to the SRB motor and the O-rings.

Back to STS-61C

We need to return to STS-61C for a moment to bring up the point that the flight was scrubbed on the first attempt to launch. Congressman Bill Nelson was on this flight as a payload specialist, which is another way to say passenger who did nothing but go for the ride but this was only in some cases, not all. The reason for the scrub had to do with an over-speed indication on the turbines used in the APU (auxiliary power unit) that was part of the SRB steering system. We are using this point to show something. Excess pressure was put on a launch because of a V.I.P. on the flight. Mulloy was angry that this occurred. He told crews they would have to work over the Christmas vacation to fix the problem.

Mulloy took considerable heat from the LEVEL I and II shuttle program team and with Dr. Lucas running Marshall, you can be assured that Mulloy was seriously chewed on for the flight delay. It did turn out that the overspend indication was actually caused by a burned sensor, a part that had come from a new vendor, and this was the first time a part was used from them. Due to all the stress that Mulloy took for just about everyone else in the shuttle program at Marshall, they would need to think twice before a launch delay would be called due to a mechanical malfunction. Why? Because it was part of the NASA pecking order, anyone (center) responsible for a shuttle delay would have to answer to his or her director. In this case, the dragon in the den was Dr. William Lucas, who already had quite a reputation for blowing fire on his workers that didn't comply with his rigid standards. Of course, we have to look at the NASA Administrator's office at this junction, basically because there was no control from the front office. Dr. Graham was a newbie on the block, and he wasn't about to countermand a fellow like Dr. Lucas who had the credentials to hold his position.

Thiokol was in an uproar because NASA chose not to sign the contract for the "BUY III" proposal. NASA was still looking for a second source for the SRB and that was official. Thus, we have pressure on the side of one of the main contractors that was also adding oil to the fire. Consider the date of STS-61C (Columbia-January 12, 1986—24th flight) just days before the ill-fated Challenger flight.

Already late for a sensor malfunction, a Congressman on board and late to the launch, it didn't look good. The next flight, number 25, STS-51 L had the first teacher in space. NASA couldn't lose face again. The press, Congress and the public wouldn't have it.

Coming down to the day

Here we are coming down to launch day for STS-51L. Let's take the pulse of what was happening at NASA and Thiokol. Thiokol got the shock of its life one week before the scheduled launch of STS-51L. NASA formally announced that it was going to go ahead with a plan to seek bids for a second source of production for the SRBs. Larry Mulloy told Thiokol he would be chairing a meeting at KSC to discuss the possibilities of giving total responsibility of SRB disassembly in Hangar AF, from Thiokol to USBI (United Space Booster Inc). The word was out that Mulloy was best friends with George Murphy, the president of USBI and Mulloy had spent some very considerable time staying at Murphy's Florida home when he was at KSC for launch activities. It was not secret that many of NASA's former employees were now working for USBI. That didn't help Thiokol's cause one bit. There were also some issues out there concerning the solvency of USBI and that Mulloy was trying to help them bail out of their financial problems. If Thiokol management had not felt the SRB work would be second sourced out to another company, possibly a top issue of the Challenger disaster may not have happened.

The Shuttle Enterprise separating from the SCA in test trials (NASA)

There was nothing to compare with the majesty and power of a shuttle launch (NASA)

The shuttle Atlantis on the back of her SCA (747) on the way home from Edwards AFB landing to Kennedy Space Center, Florida (NASA Carla Thomas)

The SCA in a refueling exercise sans the shuttle with a NASA T-38 in attendance
(NASA)

The burned out cabin of Apollo AS-204 that killed astronauts,
Ed White, Gus Grissom and Roger Chaffee.
(NASA)

The ravages of the fire on Apollo 204 are obvious (NASA)

The NASA meatball logo of the 1960s (NASA)

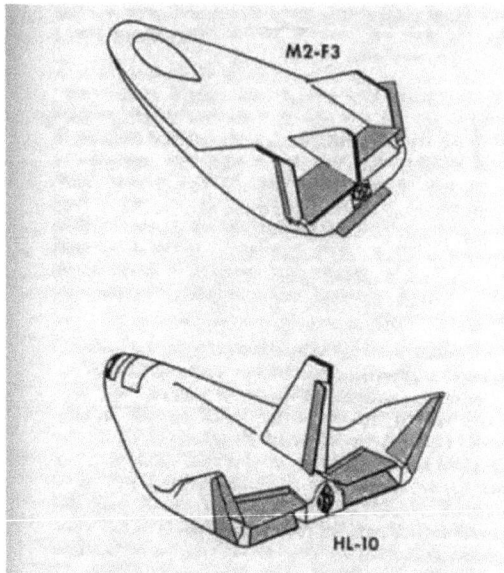

The M2-F3 and the HL-10 lifting bodies that led to the shuttle planform (NASA)

The X-15 Rocket plane led to much of the research for the Space Shuttle
and many more aircraft (NASA)

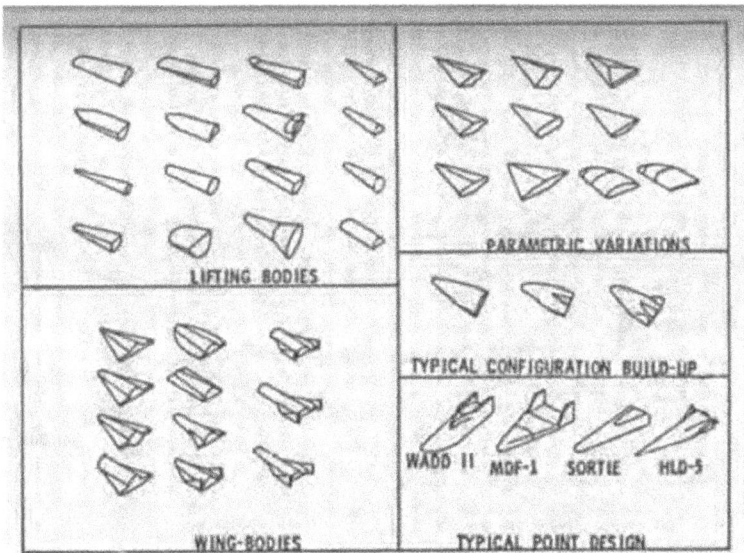

The many different planforms for lifting bodies involved in research
for the shuttle and the space plane (NASA)

The early General Dynamics "Triamese" concept (General Dynamics)

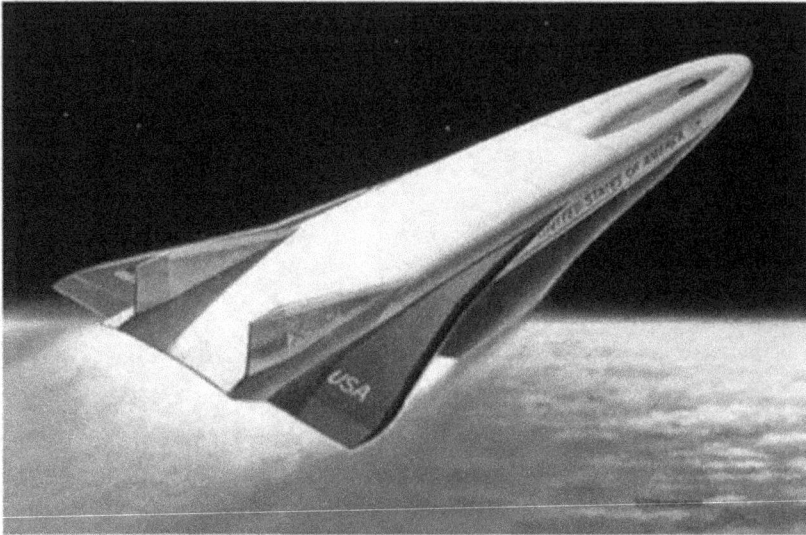

The NASA "Space Plane" concept by Rockwell that didn't get far (NASA)
USAF's Vandenberg launch site in California (USAF)

USAF's Vandenberg launch site in California (USAF)

OV -102 shuttle Columbia turning into the VAB after a trip home to KSC
(NASA/Landis)

Assembly of the shuttle Endeavour at the VAB Kennedy Space Center
(NASA /Landis)

1980 cockpit of the Shuttle, looking rather dated for a spacecraft. These cockpits on the Shuttles were later brought up to speed as all "glass cockpits" (NASA /T Landis)

Working on the underside tiles of the Shuttle. Each tile was cut to size and had a serial number for Placement. (NASA/T Landis)

Construction and stacking process of the SRB which
was done in sections (NASA/ T Landis)

The section of the SRB showing the rubber like solid fuel
propellant about to be attached (NASA)

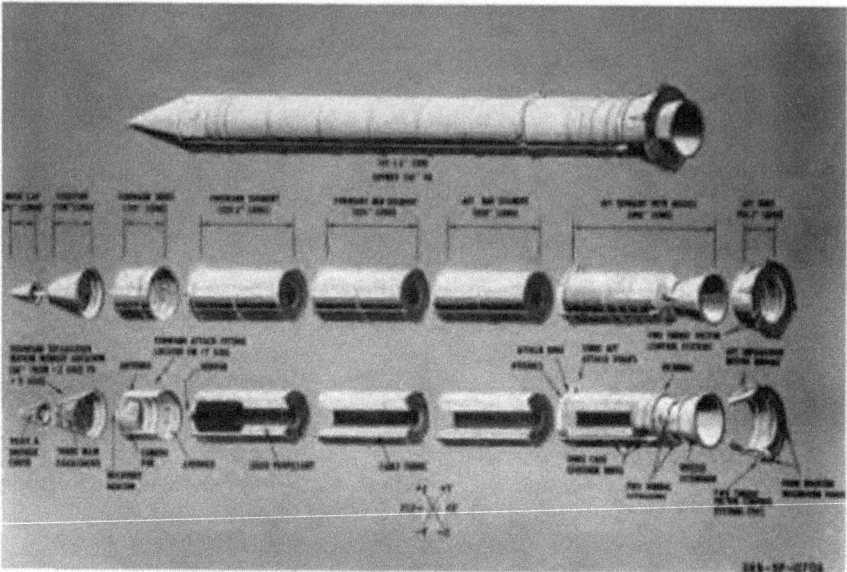

Breakdown of the SRB (Morton /Thiokol)

Image from NASA'S cameras showing the puff of black smoke coming from the
SRB of Challenger at launch, January 28, 1986 (NASA)

Left to right: Challenger crew STS-51L Onizuka, McAuliffe, Jarvis,
Resnick, Scobee (Cmdr), Smith, McNair (NASA)

The sadly famous photo of the Challenger, 73 seconds after launch with that
interminable "Y" forming as the SRBs were separated (NASA)

The ice on the pad the morning of January 28, 1986, (NASA)

Frame from NASA video showing the impact of the foam hit on
Columbia during takeoff. (NASA)

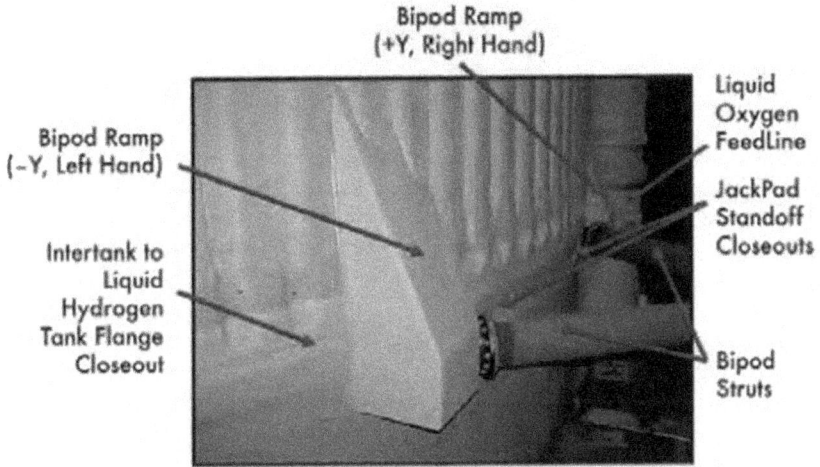

Bipod Ramp
(+Y, Right Hand)

Bipod Ramp
(-Y, Left Hand)

Intertank to
Liquid
Hydrogen
Tank Flange
Closeout

Liquid
Oxygen
FeedLine

JackPad
Standoff
Closeouts

Bipod
Struts

The bipod ramp where the foam broke off from
to strike the left wing (NASA)

The final minutes before the launch of STS-113 with Discovery (NASA)

Moving to the launch pad on the crawler, it took hours to get this done. (NASA)

STS-107 on the morning of January 16, 2003. This was the last launch of the shuttle Columbia (NASA)

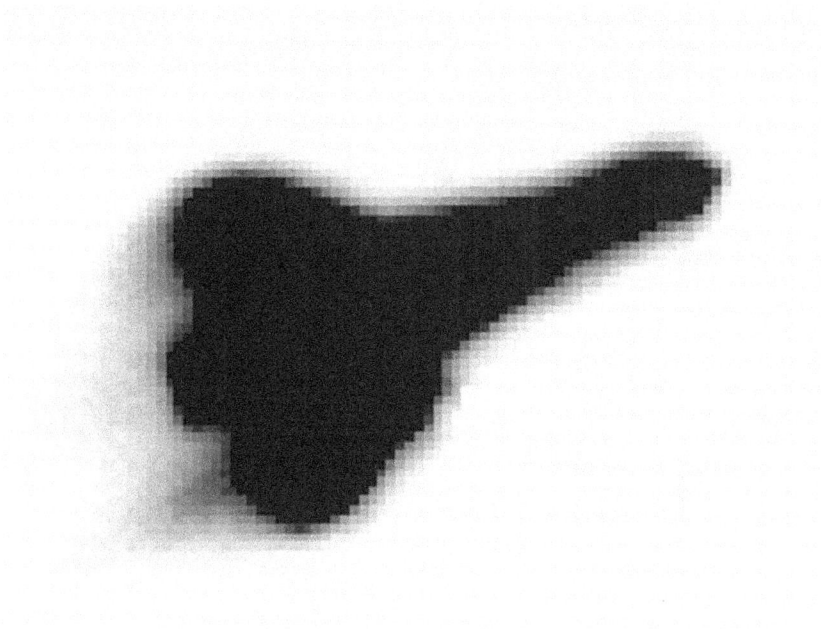

The eerie image caught by a telescope of the shuttle
Columbia breaking up on re-entry over Texas February 1, 2003(NASA)

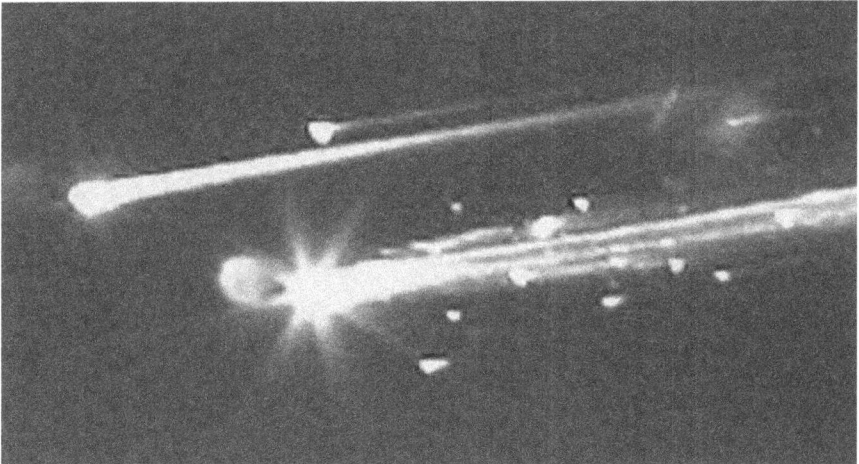

The breakup of the shuttle Columbia on reentry
as seen by hundreds of people over Texas (NASA)

The crew of the STS-107 left to right: first row Rick Husband,
Cmdr, Wiliam Mc Cool pilot, back row. David Brown,
Laurel Clark, Kalpana Chawla, Michael Anderson, Ilan Ramon.
(NASA)

The ORION spacecraft ready to go on the test stand (NASA)

ORION on her first ride to space in test 2014 (NASA)

ORION being recovered after a successful
landing by USS Anchorage (NASA)

More politics

It was not secret to anyone that the Reagan administration had developed a "Shuttle Only" policy. What that meant was the Space shuttle would be the ONLY vehicle used in US space launches. The issue here was that since NASA could show that they could launch two shuttles a month, under great internal duress however, the CELV (complimentary expendable launch vehicle) was pushed through Congress and that was a threat to the shuttle program. This was great for the Air Force; they finally got something they wanted. However, once the CELV program was on the docket, the USAF could withdraw all the satellite launches that were lined up for the shuttle. That wasn't good for NASA or the shuttle program. NASA knew they were in deep waters, if the USAF pulled out they had a void the size of the Andromeda galaxy to fill for work and money.

As we return to KSC, STS-61C finally launched after seven attempts. The Press was all over NASA for the delays and the pressure was just ready to go over the rim. This particular launch was delayed more than any other launch on the books.

The press was haranguing NASA about their two shuttles a month launch plans, when it took nearly a month to launch one. The STS-51L launch was the next and the January weather wasn't looking too good.

How not to launch a shuttle

STS-51L was on the launch pad 39B. It was the first time that this launch pad was used. January in Florida was usually mild, but there were days that a cold snap could make you feel that you were actually in New York City. The weather during Challenger's launch window was so tricky that several launch attempts were made between the original date of January 22nd and January 28th.

There were some that were happy it was delayed, because of Super Bowl Sunday, which fell on the 26th. That was a shame because the weather that day was superior. Another delay was on the books because the Orbiter's crew compartment door. On latching it, a pad technician had tried to remove the handle from the door, as was checkout policy. He couldn't get it off. Lockheed engineers tried with different tools to detach it. There was a problem with the tools. Most of them were battery driven and had not been charged to full capacity. Hence, they would run down in mid procedure. This scene was NASA's worst nightmare. We weren't talking about a delicate, mechanical part, this was a door handle! In a final effort, it was hack sawed off, allowing the shuttle to launch.

There was a weather launch window to contend with and that was closed due to the handle problem. The mission would be scrubbed until the next day.

The meeting to end all meetings: the decision to launch STS-51L

As mentioned, Florida weather was an iffy issue in January. There was a cold front due the weatherman said, and he said it would be an "extreme cold front". The weather team said that freezing would occur around midnight with at least 22 degrees expected at 6AM and possibly going up to 26 degrees by 9:30 launch time. Thiokol knew they were going to have a problem with the O-rings. The teleconferences, for which NASA was famous, would start. The connect went like this: NASA /Marshall to Kennedy Space Center to Thiokol, Utah. Thiokol wanted Bob Lund (their VP of engineering) to be deeply involved in the engineering decision, NOT in program management. There was the fine line. Thiokol called NASA resident manager at Marshall and KSC to let them know of the concerns of sealing abilities of the field joint O-rings with temperatures in the teens. They asked for the teleconference. The link was set up with Thiokol, Utah and Marshall in Huntsville and KSC in

Florida. The people involved, Larry Mulloy and Stan Reinhartz (shuttle project manager) for Marshall. The meeting was set for 8:15 PM so the decision could be made before fueling on the external tank would begin at 12 AM.

The meeting began with charts and information being faxed back and forth by 8:30PM. We all need to remember that in the 1980s, this was state of the art in communications. It was close to 8:45PM before the teleconference began and even then, Thiokol had not received any charts on the conclusions and recommendations. Since this conference was such a last minute affair, the engineers at Thiokol in Utah didn't have time to come up with any charts, they pretty much hand wrote their concepts and took some from an earlier presentation made in August 19th on the erosion of the SRM pressure seals that was done at NASA headquarters in Washington, D.C.

9:15 PM rolled around before all the charts with conclusions and recommendations were finally ready. Thiokol's presentation added up to twelve charts of engineering analysis of the joints. Half of the charts were created by Roger Boisjoly, the head seal expert from the O-ring task force. The first charts showed a history of O-rings erosion and the bow by seen in the field joints of prior shuttle flights. The next presentation recalled what was said at the August 19 conference concerning the inability of the second seal to maintain pressure during motor operation should the primary seal blow out due to erosion and blow-by.

Boisjoly had spent monumental time pointing out the timing functions of the O-ring primary seal functions, which was extremely important during the ignition phase of lift off. He said[16]; *"With colder temperatures we're moving away from the direction of goodness…. to*

---

[16] "Report on the Presidential Commission of Space Shuttle Challenger Accident June 6, 1986

where the sealing capability of the primary O-ring failed during the last half of ignition transient, a period lasting from 0.33 to 0.600 seconds" Boisjoly went on to show an exaggerated pressurized field joint, that clearly showed how the secondary seal would not be in contact with the tang sealing surface; while the primary O-ring was extruded by the motor pressure into the gap between the two seals. Another August 19th point made to NASA headquarters presentation found Boisjoly saying how the loss of the seal contact would result from joint rotation, *"The point I am making is that the primary O-ring fails to seal before the joint rotates, then yes, the secondary joint rotates and the primary seal is lost. The secondary seal has a high probability of sealing the joint. However, if the joint rotates and the primary seal is lost, then the secondary seal will be unable to seal the joints. That's why the timing function for the dynamics seal is so important, gentlemen."* Boisjoly felt so strongly that cold temperatures would slow down the timing mechanism and make it more difficult for the primary O-ring to seal the joints. He went on to show another chart on how cold temperatures would reduce all the factors that helped maintain a good seal in the joint lower O-ring squeeze due to thermal shrinkage of the O-ring. Thicker and more viscous grease around the O-ring made it slower to move around the O-ring groove and a higher O-ring hardness due to colder temperatures made it more difficult for the O-ring to extrude vigorously. The actuation or timing functions of the O-ring when at the very same time the O-ring could be eroding, creating a situation which would result in the secondary seal not being able to seal the motor. That was if the primary O-ring was sufficiently eroded to prevent sealing in the joint.

The chart of STS 51C showed large areas of jet black grease from the seal between the O-ring in one field joint on each booster including a temporary loss of the primary seal at ignition. On the STS 51C flight on January 1985 was the coldest launch to date and Boisjoly was sure that conditions seen in STS 51C was *"A direct function of the cold tempera-ture on that launch."* What Boisjoly was referring to had to do with the material the O-ring was made of. The product was called Viton. It wasn't a new material; it had been used in many situations before in the aero-space industry. Viton was made of fluorocarbon and it got harder when exposed to a low temperature, which in the case of the O-ring in the shuttle, would cause it to be harder when pressure was actuated and it extruded the O-rings into the seal gap. Boisjoly said, *"The cold tempera-ture predicted can change the O-ring material from hard sponge to something more like a brick.,"*

Arnie Thompson was the structure manager for Thiokol showed that the test data on resiliency of the O-ring on dynamic blow-by and on the compression of an O-ring. Resiliency was the ability of the material to spring back to its original shape after being compressed or squeezed. It was important for an O-ring between two plates and this rapidly moving away in one of the two plates. The test examined how long it took the O-ring to spring back and contract the other plate without pressure assis-tance. The data showed that at 100 degrees, the O-ring would never lose contact with the plate but at 50 degrees, it lost contact wand took some 10 minutes to recover fully. That is a frightening scenario, yet it was scary enough for NASA. The cold temperatures could dramatically reduce the ability of the O-rings to seat on the surface that was moving away from each other like the field joint. This chart showed the worst-case scenario as was see on STS 51C in January 1985.

Both the case and nozzle joints were affected. Yet in another chart, we see blow-by in field joints of only one other flight, in which two joints of the left hand booster of STS-61A, during the warmest launch days in October 1985 occurred. When this chart was shown to Larry Mulloy, he challenged with, *"How then can you conclude that temperature has anything to do with blow-by?"* Roger Boisjoly replied with; *"We need to explain the difference between that the data obtained in STS-61A in October and that observed on STS-51C in January."* Boisjoly personally inspected the SRM motors on STS-51C when they were dissembled at the KSC facilities, Hanger AF. Someone else had inspected the STS-61C motors. Boisjoly saw that both joints of the 51C flight had primary O-rings erosion. He also saw a quantity of jet-black soot over a large portion of the circumference between the primary and secondary O-ring from 80 degrees to 110 degrees.

The argument continues

The STS-61 C joints on the warmer launch day had no primary O-ring erosion and only a very light gray coating of soot over the small circumference between the O-rings.

This test observed the difference between the two flights was due to temperature ONLY.

Boisjoly commented, *"That the observed differences between these two flights is most probably due only to temperature—that STS 51 C had O-ring temperatures of 53 degrees and STS 61 C had O-ring temperatures of 75 degrees."* However, Mulloy wasn't buying that. NASA had launched shuttles at various times and temperatures between 53 degrees and 75 degrees without seeing any blow-by or field joint issues. Mulloy refused to believe that Thiokol could conclude that temperature had anything to do with the issues at hand. He felt the data was inconclusive. Boisjoly kept fighting and disagreed. He showed his last chart that compared the O-ring squeeze of the primary seals used in STS 51C. The chart showed the magnitude of the primary O-ring erosion on the STS 51 C blow-by. The chart went on to show the differences in cold temperatures versus warm temperatures, but no matter how you sliced it, O-rings reacted slowly and became harder in temperatures 53 degrees and below. That didn't matter to Larry Mulloy. He didn't give an inch on the concept that temperatures made no difference whatsoever. Boisjoly also wouldn't budge on his charts and what they showed. Cold temperatures made a major difference in the way the part reacted, period. This meeting was already an hour long and it was 10:00 PM before the final touches were put in.

NASA continued to disagree, and Thiokol continued to push the temperature plan. The last chart put the icing on the cake---temperature was not the only parameter that controlled the blow-by that was seen on 51C at O-ring temperature 53 degrees. The conclusion: at 50 degrees blowby could be seen in case joints and the temperatures for STS 51 L on January 28, 1986, would be 29 degrees at 9:00AM and 38 degrees by 2:00 PM. There was no difference in either motor used other than the temperatures.

The chart looked something like this:

CONCLUSIONS:

- Temperature of O-ring is not only parameter controlling blow by.

- SRM 15 (STS-51C) with blow by had O-ring temperature of 53 degrees.

- Four development motors with no blow by were tested and the O-ring temperatures were 47 degrees to 52 degrees F.

- Development motors had putty packing which resulted in better performance

- At about 50 degrees F blow by could be expected in case joint

- Temperature for STS-51 L on 1/28/86 will be 29 degrees at 9:00 AM and 38 degrees at 2:00 PM

- Have no data that would indicate the 51 L motor is different from other SRMs other than temperature.

RECOMMENDATIONS:

- O-ring temperature must be > (greater than) 50 degrees.

- Development motors 47 degrees to 52 degrees with putty packing had no blow by.

- Project AMBIENT condition temperature and window to determine launch time.

Mulloy still would not budge. He wouldn't accept the rationale used to arrive at the recommendations. Stan Reinhartz then asked George Hardy, deputy director of Science/engineering-NASA/Marshall for his opinion. Hardy was no better in his determination. In fact, he was. Aghast that Thiokol could even bring these recommendations to the table. However, he was determined that he wouldn't want to fly without Thiokol's agreement. Hardy felt that Thiokol was only addressing the O-ring issue and did not answer the question of whether the secondary O-ring would hold up. Mulloy literally shouted; *"My God, Thiokol, when do you want me to launch? Next April?"*[17] These words went down in history as not so much a knee jerk reaction, but fateful words, cold and stubborn. Mulloy also added "the eve of a launch is a helluva time to be generating new launch commit criteria". In fact, Thiokol was not trying to add a new launch commit criteria, they were trying to point out that there were things that change as the Shuttle program grew. NASA refused to see that and refused to give the leeway needed to make those changes. Because of the bad press, pressure to prove the shuttle program was viable, the stubbornness and lack of foresight to see that engineering is NOT a perfect science led to the death of seven of our best astronauts and an Orbiter.

---

[17] The Challenger Launch Decision - Risky technology, culture, and deviance at NASA - Diane Vaughn

The meeting continued with Stan Reinhartz adding that he was under the impression that the SRM were qualified for between 40 degrees and 90 degrees. The 50-degree recommendation was not consistent with that. Larry Mulloy asked Joe Kilminster (VP of Space Booster Program – Thiokol) what the program office recommendation was and Kilminster said he would not recommend the launch based on the engineering just presented. Mulloy was still unhappy with the engineering position that Thiokol just presented. Based on Mulloy's assessment, the engineering data was not conclusive. Mulloy felt that Thiokol presented the data that observed blow-by on cold motors and warm motors. Mulloy wanted more quantitive data that temperature really affected the ability of the joint to seal. Mulloy wasn't giving up and he was adamant to challenge this diagnosis. He was not going to accept the recommendation that was just made based on the data presented. Stan Reinhartz, then asked Kilminster to respond to Mulloy's remarks. He couldn't. Kilminster then asked for a five minute off line conference to reevaluate the data, which Reinhartz agreed to. Thiokol went off line to talk.

Thiokol and the reevaluation

Thiokol left the teleconference to talk off line. They were going to re-evaluate the data. They needed to consider a comment made earlier by NASA's George Hardy that was relative to the ability of the O-ring to seal because no one on Thiokol's team responded to the comment.

Lower temperatures were rough for both the O-ring and the timing mechanism. It slowed it down. That effect is much worse for the primary O-ring into the wrong side of the groove while the secondary O-ring goes in the right direction. This condition should be considered in the final decision for recommending the lowest acceptable temperature for launch. Thiokol's last chart was the only chart shown in defense of Thiokol's theory, of which Mulloy was chewing on. Thiokol felt the O-ring was the biggest part of the problem. It should be engineering decision not a program manager decision to launch or not. However, that isn't how it turned out.

Thiokol was still worried about the STS 51C flight of 1985 when an O-ring lost its seal during and early phase of the ignition sequence and that is why the secondary O-ring worked well as the redundant seal but had some minor damage.

Teleconferences and decisions

The main concept here had to do with temperatures. At what temperature did the O-ring fail to seal properly and how did temperature affect the timing mechanism with cold temperatures slowing it down. What was the lowest temperature at which the primary and secondary O-rings were no longer viable, because the timing function was halted enough that Thiokol was truly worried about the problems.

The secondary seal caused NASA to pull it down from a *Criticality 1* redundancy to a **Crit 1** in late 1982. NASA people at the meeting were callous at the teleconference, especially Larry Mulloy. Mulloy was challenging Thiokol on the justification that it was safe to fly. The first time NASA ever challenged a recommendation from a contractor that it was unsafe to fly. Thiokol was in the position to prove the seals would definitely fail and Thiokol just didn't have the time to do the "charts" to prove it. Mulloy and his team weren't ready to accept a change to flight criteria if Thiokol stood on their heads. NASA managers reasoned that they were right for not accepting the any launch recommendation from Thiokol. They felt that Thiokol's data was "inconclusive and not quantifiable". The strange thing was in the past; any item that would come up as inconclusive or unquantifiable would be an automatic consideration to not launch. What changed?

The NASA Marshall team put a lot of pressure of Thiokol, mainly their VP of engineering, Joe Kilminister to change his mind and support a decision to go with a shuttle launch. On the other hand, Larry Mulloy ignored Thiokol's VP of Space Boosters Kilminster's recommendation. Mulloy asked Kilminster what he thought. Kilminster stayed on the side of Thiokol and the decision to stand down on this launch. Mulloy had one heavy ace up his sleeve and that was with Kilminster. Basically, Kilminister worked for Mulloy and was his program management executive counterpart at Thiokol. Thiokol was behind schedule on its current program, and didn't have a signed contract from NASA for the next only source procurement of 66 sets of motors; Kilminster was in a difficult situation. It was getting late and Mulloy wanted a signed recommendation by early morning of the next day, launch day.

Boisjoly too, was frustrated and deeply concerned. He produced photographs of the soot that was seen between the primary and secondary O-rings on STS 51C. Boisjoly was feeling this so deeply.

"Look carefully at these photographs! Don't ignore what they are telling us, namely that low temperature causes more blow by in the joint."[18] The looks he got from across the table was a "deer in headlights" stare. At that point, Boisjoly just sat down. No one in the Thiokol team said yea or nay because they didn't want to go against management. The same material was rolling around and around.

It was getting everyone nowhere in a hurry. Jerry Mason, who was the General Manager and senior VP of the Thiokol Wasatch group, nailed Bob Lund, VP of Engineering for Thiokol to the floor. Mason told Lund, who was silent to the ongoing tug of war between NASA and Thiokol, "It's time for you, Bob, to take off your engineering hat and put on your management hat." Deadlier words could not have been spoken. It was

---

[18] Presidential Commission of Space Shutttle Challenger Accident June 6, 1986

exactly what the Thiokol engineers didn't want, a management decision. They wanted an engineering decision. Lund was not only bullied, but also forced into making a decision that he would have to live with forever. Lund agreed that the launch should go off as scheduled. The teleconference was over and all that was needed was the signed paperwork, which NASA was waiting for the next morning. As far as NASA was concerned, the O-rings were certified for 40 degrees F and that was okay with them.

However, there were people at Thiokol who were dreading the next morning's launch. The fax was sent that evening to NASA, signed by Joe Kilminster, VP of Space Booster Program for Thiokol.

MTI ASSESSMENT OF TEMPERATURE CONCERN ON SRM-25 (51L) LAUNCH

- o CALCULATIONS SHOW THAT SRM-25 O-RINGS WILL BE 20° COLDER THAN SRM-15 O-RINGS

- o TEMPERATURE DATA NOT CONCLUSIVE ON PREDICTING PRIMARY O-RING BLOW-BY

- o ENGINEERING ASSESSMENT IS THAT:

  - o COLDER O-RINGS WILL HAVE INCREASED EFFECTIVE DUROMETER ("HARDER")

  - o "HARDER" O-RINGS WILL TAKE LONGER TO "SEAT"

    - o MORE GAS MAY PASS PRIMARY O-RING BEFORE THE PRIMARY SEAL SEATS (RELATIVE TO SRM-15)

      - o DEMONSTRATED SEALING THRESHOLD IS 3 TIMES GREATER THAN 0.038" EROSION EXPERIENCED ON SRM-15

  - o IF THE PRIMARY SEAL DOES NOT SEAT, THE SECONDARY SEAL WILL SEAT

    - o PRESSURE WILL GET TO SECONDARY SEAL BEFORE THE METAL PARTS ROTATE

      - o O-RING PRESSURE LEAK CHECK PLACES SECONDARY SEAL IN OUTBOARD POSITION WHICH MINIMIZES SEALING TIME

- o MTI RECOMMENDS STS-51L LAUNCH PROCEED ON 28 JANUARY 1986

  - o SRM-25 WILL NOT BE SIGNIFICANTLY DIFFERENT FROM SRM-15

(signature)

JOE C. KILMINSTER, VICE PRESIDENT
SPACE BOOSTER PROGRAMS

MORTON THIOKOL INC.
Wasatch Division

[Ref. 2-28-6]

This fax was sent to NASA the evening of the launch of Challenger.
It was actually the death warrant for the Challenger and her crew.

January 28, 1986

The dawn at Kennedy Space Center was bright, blue and freezing cold on January 28, 1986. Many of the contractors, workers, bystanders, saw in those early morning hours how the ice had formed on launch pad 39B. Teams were sent out by NASA to assess the quantity of ice that had formed around and on the launch pad.

The sound suppression system that also used enormous amounts of water, was also was frozen. The temperature between 1:30 and 3:00AM was around 29 degrees. There were large quantities of ice on the FSS (fixed service structure), the mobile launch platform and the launch pad apron. The water trough underneath the SRBs was filled with antifreeze that was supposed to be active up to 16 degrees but ice sheets were already forming. Because of the cold, some of the water lines were left running to prevent them from bursting. Three inches of ice had formed around the pipes.

To top it all off, there was an ocean breeze coming in, blowing water all over the launch facility.

The second team that went out to assess the launch site, saw temperatures of 26 to 30 degrees F. Ice had gotten thicker around the water troughs below the SRBs. Actually, there were now sheets of ice that formed even with the antifreeze. Icicles were hanging from every available spot, but most importantly there were icicles on the left hand SRB and its aft skirt, including on the left hand case and skirt. Icicles were removed, but the sheets of ice were left. Why? The word from the head of the Ice Team for NASA, Charles Richardson told the director of NASA Engineering at KSC, Horace Lamberth. The only choice you got today is not to go." More measurements taken of the left hand SRB and the aft skirt proved to be around 26 degrees. The right hand booster read somewhere around 9 degrees and even lower on the right hand aft skirt, 7 degrees. The water trough was measured between 8 and 10 degrees, while the mobile launch platform was 12 degrees. The Ice Team gave a briefing to the Mission Management team just after 9:00AM. We should stop right here for a moment. I am sure that everyone reading this book has a good deal of common sense. I am sure that if you saw conditions like this on your car and roadway before a trip, you would consider staying at home. In fact, I know that many of you have done it, as I have. Where in the name of common sense, forget about engineering, NASA politics, Thiokol's bottom line, would anyone in their right mind (I guess that is the operative word here) decide to launch a complex machine like the shuttle that is carrying human life in conditions such as these. I believe sincerely that any reasonably educated person would have looked at this scene and said, "Wait". However, the common sense that was lacking or misplaced at NASA was nowhere to be found.

Directors from NASA, engineers, technicians, you name it, all saw the same thing.... ice and cold, deep ice and cold. We are not talking about 48 degrees; we are talking about holding a machine out in freezing overnight temperatures with parameters on parts that needed to be met. NASA just disregarded that because they had a schedule to maintain. It was a big launch today, teacher in space! Hindsight is a glorious thing and I bet the families of all the astronauts would have loved to have one voice with common sense that would have said, "Wait... Scrub the launch for one day." Yet, that didn't happen, as we all know. The result of all of NASA's engineering skills, program management, technical expertise, went up in a ball of horrific flames at 73 seconds into launch. "Go at throttle up." The next thing we saw were smoke columns forming huge "Y". We ask the same thing. The answer lies in the teleconference of the night before. The tug of war with NASA/Marshall and Thiokol was truly a study in stubbornness on the part of NASA and weakness on the part of Thiokol. What did it all boil down to? That's easy, money for Thiokol—the bottom line and saving face for NASA—they just couldn't' have the press badgering them for being late on a launch again. Having the teacher in space program and the world looking at another scrubbed launch would have been too much. You can try to balance this out with all kinds of sociological arguments that have to do with problems starting at the base line and working their way up for years. You can try to piece together the remarkable stories of Alan McDonald and Roger Boisjoly, both Thiokol engineers.

You can read other depictions of this event until the cows come home, but what it all comes down to is this, common sense and the guts to buck the system and say no. Some tried and failed because they were not high enough up in the food chain to make a difference. Others remained silent and went with the flow. Others decided that there wasn't enough of a "critical issue" according to the rules and regulations that NASA had developed and others had to interpret. However, the bottom line is this, seven wonderful, talented people and one beautiful machine were destroyed because of arrogance, rigidity in thought processes when dealing with engineering principles and political/monetary avarice. That is the reason, pure and simple. When you boil down the information out there, this is what it all comes to. It's a terrifying thought, even now so many years later. Now, that we no longer have a manned space program worthy of the name.

73 Seconds to Eternity

Here we are at the launch pad 39B, getting ready to send Challenger off on her mission. There had been several launch attempts starting from January 22 right up to January 28. This was the first mission to be launched off Pad 39B. The objectives for Flight 51-L were to deploy the Tracking and Data satellite (TDRS-B) and to fly the Spartan Halley experiments. Spartan was to deploy from Challenger's payload bay and send two ultraviolet spectrographs to the tail of the Halley comet. The mission also included the much-publicized Teacher – In – Space, Christa McAuliffe. She was to do two classes in space.

The shuttle launched much like any other, except the ground temperature was 36 degrees F. This was actually 15 degrees colder than any other previous launch. Of course, we know about the fights and issues of the night before. Challenger lifted off at 11:38 AM. At approximately 37 seconds, Challenger felt the first of several high altitude wind shear conditions that rocked her. It lasted somewhere around 64 seconds. The wind shear created forced on the shuttle with violent fluctuations. The shuttle sensed the condition and immediately compensated for it with the guidance-navigation-control system. The loads placed on the orbiter exceeded the prior experience in both yaw and pitch at certain times during the liftoff. The steering system of the SRB had already responded to all commands and wind shear effects. These wind shear effects caused the steering system of the Orbiter to be much more active than any other flight so far.

Some 45 seconds into the flight there were three bright flashes that appeared just downstream of the shuttle's right wing. Each flash had lasted less than a 1/30 of a second and was caught by the NASA stationary cameras. However, flashes like this were seen on other flights. Another appearance of a separate bright spot was see by the film analysis that seemed to be a reflection of the main engine exhaust on the Orbital Maneuvering System pods, OMS pods that are located at the upper rear section of the Orbiter. The flashes were not related to the later appearance of the flame plume that ejected from the right SRB. The shuttle's main engines and SRBs were operating at a reduced thrust because of passing through the MAX Q, or the maximum dynamic pressure of 720 psft. We need to remember at this point the main engines had been throttled up to 104% thrust and the SRBs were then increasing their thrust when the first of the flames appeared on the right hand SRB aft field joint. At something like 58 seconds, the flame was seen on some enhanced imaging taken of the flight by NASA cameras. In a later NASA camera, film the flame was clearly visible without any enhancement and literally continued to grow in centimeters into a well-defined flame by 59 seconds into the flight, past all hope of saving the shuttle or the crew.

Around 60 seconds later, the telemetry started to show in mission control that there was a serious pressure difference between the chamber pressure in the right and left SRB. The right SRB chamber pressure was definitely lower showing that the leak in the right aft field joint was growing. As the flame grew, it was deflected by the aerodynamics of the slipstream towards the rear of the vehicle. It was also pushed circumferentially by the protruding attachments of the booster to the external tank. This caused the flame to push on to the surface of the external tank. The flame was spreading and impinged on the strut that attached the SRB to the ET.

At some 62 seconds into the doomed flight, the control systems started to react to counter the forces caused by the flame. Lack of pressure in the right SRB and its effect. The left SRB thrust vector control moved to counter the yaw, which was caused, by the reduced thrust of the right SRB that was leaking. Within the next 9 seconds, the space shuttle control systems were pushing to correct all the abnormalities it was facing in the pitch and yaw rates of the vehicle.

The first indications of the flame from the right SRB ruptured the external tank at sixty-four seconds into the flight. Challenger and her crew had only another eleven seconds of life left. Seventy -two seconds on the in-flight clock and the next events occurred rather quickly. The telemetry data showed that a very wide number of flight system reactions occurred that supported the visual scene of the photos as the Orbiter struggled against all the forces set to play against her. They were already tearing her apart. At seventy -two seconds into the flight the lower strut, linking the SRB to the External Tank was either cut or pulled away from the already weak hydrogen tank. That permitted the right SRB to rotate around the upper strut. The rotation was showing deviating pitch and yaw information between the right and left SRB.

We are now at 73 seconds into the flight. A white vapor pattern was seen developing from the side of the external tank bottom. This was the end for the Orbiter and the External tank. The structural failure of the external tank created a sudden forward thrust of about 2.8 million lbs., which pushed the External tank into the intertank structure. At this same time, the rotating right SRB impacted into the intertank structure and the lower part of the liquid oxygen tank. The structure began to fail at 73.137 seconds, which we all remember as the wide, white smoke, Y formation. In milliseconds, there was an enormous, explosive burning of the hydrogen that was streaming from the ruptured external tank. The entire vehicle was moving at something like Mach 1.92 and 46,000 feet, as the RCS (reaction control system) ruptured and the hypergolic burn of propellants started. The reddish brown color of the hypergolic fuel that was burning was visible along the edge of the main fireball. The Orbiter under severe aerodynamic loads and torque broke into several pieces, which later emerged, from the fireball. Separate sections that were identified on the NASA camera film included the main engine, tail section and the still burning right wing of the orbiter. The forward fuselage, which trailed a mass of umbilical lines, pulled away from the payload bay headed for the bottom of the Atlantic Ocean.

The Astronauts

All wondered whether the astronauts of the Challenger flight were conscious during the descent to the Atlantic Ocean. The recovered crew cabin showed that it might have been possible that the crew did not have decompression since the windows were still intact. Some or all of the astronauts might have been alive and conscious all the way to the ocean impact. The cabin did hit the ocean at something like 200 mph on the commander, Dick Scobee's side of the cabin. The Roger's Commission report said, "The internal crew module components recovered were crushed and distorted, but showed no evidence of heat or fire. A general consistency among the components was a shear deformation from the top of the components toward the +Y (to the right) direction from force acting from the left. Components crushed or sheared in the above manner included the avionics boxes from all three avionics bays, crew lockers, instrument panels and the seat frames from the commander and the pilot. The more extensive and heavier crush damage appeared on components nearer the upper left side of the crew module. The magnitude and direction of the crush damage indicated that the module was nose down and in a steep left bank attitude when it hit the water." The Commission continues, "The fact that pieces of the forward fuselage upper shell were recovered with the crew module, indicates that the upper shell remained attached to the crew module until water impact. Pieces of the upper forward fuselage shell recovered or found with the crew module included cockpit window frames, the ingress/egress hatch, structure around the hatch frame and pieces of the left and right sides. The window glass from the windows, including the hatch window, was fractured with only fragments of glass remaining in the frames. From the rest of the debris that was recovered the PEAPS (Personal Egress Air Packs) had been activated. However, the impact with the ocean was so violent that there are many pieces of the puzzle that will never be found or understood. While the forces during the breakup of the Orbiter weren't enough to kill the astronauts they possibly did lose consciousness due to pressure loss in the crew module. The acceleration or g forces on the cabin were anything between 12 and 20 times the force of gravity vertically. The astronauts were pushed violently down in their seats. According to Joseph Kerwin, director of Life Sciences at NASA," In two seconds, they were below four G's; in less than 10 seconds, the crew compartment was essentially in free fall. Medical analysis indicates that these accelerations are survivable, and that the probability of major injury to crew members is low." Since the cabin impacted at 200 miles an hour, it is sure that when the cabin hit the ocean surface, the astronauts perished. Challenger and her valiant crew were gone forever.

The bottom line

We have explored many of the issues that NASA had to face get the shuttle built. We will look at some of the closer timelines with the shuttle disaster. Many things added up to the final number for Challenger. The disaster did not happen essentially because of processes at the beginning of the shuttle project. Yes, some of the early shuttle project processes and money concerns had something to do with it, but the process was never complete. Things were added onto the situation until it became so tangled and overblown with bureaucracy, ego and blindness to the real mission, that Challenger's demise was inevitable. Late in the launch program, the pressure to keep the line moving was becoming almost intolerable. Just a month before the Challenger launch, STS-61C was given time off during the KSC Christmas and New Year's holidays with many delays for the launch of Columbia on this flight. Jesse Moore, head of KSC, okayed those layoffs, again delaying the launch timeline further. Yet, there was still a little wiggle room to allow some slack in the schedule. This allowed NASA's overworked and stressed workers a break. Another issue came up January 16-17 KSC meetings because of complex abort issues that could not be resolved.

KSC was way behind schedule in producing the written operational management orders, which were thick, printed volumes that hardly anyone ever really read. There was the issue of the MIA NASA administrator, James Beggs. Beggs was being investigated for misdoings while he worked at General Dynamics. His replacement William Graham had virtually NO experience in space programs. He came out of a black ops defense department program. Hence, NASA had no real leadership. Beggs took a long leave of absence of which he never returned. Pretty much all at NASA already knew that Beggs was not the greatest manager going as he was too aloof and distant. However, even Beggs did not approve of Graham's appointment.

We have the issue that the booster problems go back as early as 1983. The TPS (thermal protection system) tiles were also an issue. Difficult to manufacture and install, NASA's biggest fear was that they would fall off during the tremendous dynamic surges at liftoff. James Beggs had also promised that the stretched out NASA centers would not be "microman-aged," basically allowing them to control themselves and develop their own ego and personality. That also turned out to be a major problem in the shuttle program. This would not have been the modus operandi during the Apollo program, where there was strong managerial presence at all the centers and control came from Washington Headquarters. During the time of 1980-85, the employment levels at NASA centers were cut more than 10%. NASA scientists and engineers were hopping off the boat at a record 20%. This, of course, was the heart of the shuttle program's development. Attrition played a huge part in bringing down the NASA staffing load. Workers retired, transferred to other programs, or had a hiring freeze which allowed the program to drop even further.

As of September 26, 1979, NASA had planned to strip headquarters of all technical oversight responsibilities. This meant that headquarters would no longer be able to watch over safety and engineering status of the shuttle program and had no ability to monitor the flight trends, redesign issues or variance patterns. Headquarters would be left with only the accounting departments and the business management end including sales. In essence, headquarters became a "front office" in every sense of the word. That leaves us with no one in headquarters who knew what was going on. Another issue was treating the shuttle as operation when if fact there were still design changes that had to happen. It also meant that those design changes would be minimized. In short, not done or not completed. One of the first things that James Beggs did after taking office as NASA Administrator was, "freeze all shuttle design". Where did that leave the shuttle program? The House Science and Technology Commission that looked into the shuttle disaster, in an August 21, 1981 memo to James Beggs from the Marshall Spaceflight Center director Dr. William Lucas:

The memo said: *"I wholeheartedly agree with your statements that shuttle performance requirements and design should be frozen so that we can concentrate all efforts on bringing the system to a cost effective operational status."* Not only was that questionable as far as management went, it was damn stupid. What one has to realize here is that the shuttle was never "truly operational". At best, it was still developmental in all phases. Freezing program requirements was done to cut the shuttle's weight. The Orbiter's were coming out of the Rockwell facility in California overweight, thus the USAF wasn't very pleased by that because it was giving them fits about payload ability. Freezing the shuttle design was impossible. This was most especially urgent in dealing with the main engines, where near catastrophes had happened on shuttle missions.

To date, September 1983, NASA had tried to establish a routine launch procedure at KSC. The Lockheed Space Corporation would manage a new shuttle-processing contract to coordinate all pre-launch activities. The centralization of the shuttle operations took place at the Johnson Space Center where the Space Transportation System operations contract was awarded back to Rockwell. This contract consolidated the work of sixteen different contractors and would take care of most of JSC's activity in support of the shuttle operations, crew training and flight operations. The contract was worth $684 million which was over four years and seen as compensating Rockwell for the loss of its dominant position at KSC when the shuttle processing was given top Lockheed. We now see, there was also a lot of backbiting, fighting for positions and who was going to be top dog. The "Doctrine" of the operational Space Transportation System also found expression in

NASA's public relations approach to astronaut selection. The STS was a public relations bonanza for both NASA and of course, the White House namely the Reagan administration. The "Pomp and Glory" of manned spaceflight was back in force and NASA made certain that the right people got the right positions in the shuttle staff.

Here is a scary little piece of information: In a 1984 General Accounting Office (GAO) study found that NASA spent more than $780,000 flying some 2226 guests to witness the 12 shuttle launches. NASA used their private jets along with charter flights to fly in VIPs to the viewing stands at KSC. The early shuttle launches and landings were carried live on the network news shows and all were covered by NASA's internal TV system. However, president Ronald Reagan didn't attend any of the launches because of the uncertainty that the launch would go on time. NASA wouldn't want that black eye in public relations.

Public relations were a big part of the problems with the Shuttle Program. Unlike the Mercury and Apollo programs, things were not so controlled. The program was vast and couldn't be contained easily. NASA compensated by creating other diversions for the Shuttle program, like a space camp for kids. The Aerospace community couldn't put out enough books on the new Space Shuttle and the shuttle program. Artists were flown in to do oil paintings of the launch and paid well for it. Other service art programs like the U.S Air Force are done by donation only. Posters were another part of the Public Relations show. There was nothing like a good poster with a great slogan to keep the public's interest. Schools were also employed to get science experiments that would fly on the shuttle. Even the industrial scene was exploited to send experiments into space. The teacher in space was the quintessential program however, that would mean lessons from space, actual classes from above the atmosphere. How unfortunate it never took place. When looked at from inside the NASA world, many who worked closely within

NASA knew that James Beggs had built his "Tower of Babel" on a hill of quicksand.

Further, on into 1985 a report of NASA's Aerospace Advisory Panel was quietly happy about what was going on. After all, its members served at NASA's invitation and were all paid handsomely for their trouble. Yet, its critique hit the nail on the head. The Panel advised: *"NASA management would be well advised to avoid advertising the shuttle as being operational in the airline sense when it clearly isn't.... the continuing use of the term "operational" simply compounds the unique management challenge of guiding the STS through this period of "developmental" evaluation."* It was obvious that there were pressures to turnaround times, which existed for the shuttle components besides the SRB, such as a waiver of structural inspection for the Orbiter fleet. Because of the problems with hardware cause delays for some early flights that sometimes took months before they were able to fly. In September of 1984, after the 12th shuttle mission flew, Jesse Moore (associate administrator space flight) was able to claim the program was back on schedule. This meant NASA would be shooting for at least one flight a month during 1985, even though 3800 tiles were in the process of falling off the Challenger and Discovery was damaged in a processing accident. Former Secretary of the USAF, Dr. Hans Mark, was also NASA's deputy administrator under James Beggs from 1981-84 said after Challenger's demise "NASA is a first class engineering development organization, but it was never intended to be the agency for the long term operation of the Space Transportation System."

Many other problems were internalized in the shuttle program. Changes in the flight plan were the norm. There was always the short notice to load in another satellite retrieval, additions of VIP payload specialists like senators and Congressmen, changes in the scientific experiments.

This caused uncertainty in scheduling that pointed to more effort to increase flight rates and the need to hold costs down was a joke because of the many changes that NASA had to uphold. The shuttle was a system that was chasing its tail. It would never catch it.

One of NASA's deepest kept secrets was that it couldn't keep any control over its contractors. They were running rampant over NASA. The Inspector General Office and the General Accounting Office looked into just what was going on. The accountants asked Rockwell to examine 1800 parts on Challenger which found a number of them could not be certified as properly built. In a statement in the New York Times," But the company suggested that Challenger "fly as is" because of the "high design margin" and the "non-critical function" of the parts. It suggested repeated inspections and limits on the use of the parts because of possible "crack growth". One part carried the warning, "May yield under load, but no catastrophic failure." Ask yourself the question, "Would you want to fly in an aircraft that had a statement like that stenciled on the landing gear? "NASA was asking its astronauts to do just that. In fact, the astronauts never really knew all the problems of the shuttle. The auditors also found more discrepancies with contractor fraud in price markups that were over 1100% on parts that were purchased from outside vendors and resold to NASA. Cases cited showed Rockwell illegally charged NASA for work it was actually doing for the Department of Defense!

NASA was also cited for paying Rockwell's Rocketdyne division overcharge on mundane items like $120 for a $3.80 bolt or $80 for a 41cent washer. The list goes on, right up to NASA paying for a contractor's idle, downtime, contractor employees who did nothing. That added up to some 30% of time charged out by Rockwell to NASA. Also reported were instances of narcotic drug use at Johnson Space Center, along with theft of government equipment and falsifying the quality of parts. To add to this, the contractor work done at KSC wasn't exactly top of the charts in quality. On March 8, 1985, the Orbiter Discovery took on serious damage in a processing accident that cost the shuttle schedule of flights at least a good month. A work platform slammed into the side of the Orbiter causing two large gashes in Discovery's side and breaking the leg of a technician. To add on to the package, the overtime work put in, along with worker fatigue, was causing calamities all over the KSC facility. A very tired contract worker on January 6, 1985, caused the scrub of a launch for STS-61 C because he drained some 18,000 lbs of liquid hydrogen from the external tank. Controllers noticed it via a change in the tank's temperature. Lift off with such a condition would have resulted in tank failure and a massive explosion on the pad.

At the start of the shuttle flights, NASA's safety mechanism was non-operational. To say it again, it didn't exist. It was all on paper, but none of it was transformed to on the scene workable, scenarios. There is also the very disturbing lack of a launch escape system. This was discussed in a 1979 budget hearing led by Senator William Proxmire. He asked NASA administrator, Robert Frosch, *"If you delay the shuttle further, could you fly more safely?"*

Frosch responded, "I don't think so, what I want to do is continue to look at a more stringent test and certification program, which is what we are doing. We will not fly till it's accomplished." Proxmire then read from the weekly bible for aerospace, the *"Aviation Week and Space Technology* magazine. *"The primary abort mode in the event of a pad emergency such as a propellant explosion, is to run quickly from the vehicle."* That should put a chill up your spine. Frosch's answer was even more chilling. Frosch responded to the Senator: *"It does not have an escape… we have not found any way to deal with the problem."* There was no way the NASA "run like hell escape plan" was viable. If a shuttle blew up on the pad, nothing was going to be saved. Proxmire later told the Washington Post newspaper that it was "NASA's military overtones that caused Congress to overlook safety problems." "You don't ask the Pentagon about safety." That was an understatement if there ever was one. NASA's egotistical "Can do" attitude left little any form of doubt anywhere and that was a big part of NASA's Challenger disaster. It was either yes or no. There was no gray area and that is what killed Challenger and her crew. Because of all these things that were discussed, prior to launch and after launch, and all the other issues that were going on during the decision making process, NASA failed to bring in one thing to the table, there had to be room to say, "change this." Egos needed to be left at the door. Instead of asking contractors to put on different hats to convince NASA of something, NASA's people should have known in their gut and listened to their contractors. They chose not to and seven died along with a vehicle.

Report of the Presidential Commission on the Space Shuttle Challenger Accident and what did we really learn from it?

The *"Report of the Presidential Commission on the Space Shuttle Challenger Accident"* also known as the *"Rogers Commission"*, (named for William P. Rogers, who headed the commission as chairman. The other commission members were Vice Chairman Neil Armstrong, David Acheson, Eugene Covert, Richard Feynman, Robert Hotz, Donald Kutyna, Sally Ride, Robert Rummel, Joseph Sutter, Arthur Walker, Albert Wheelon, and Chuck Yeager. Rogers was the attorney general for President Eisenhower and secretary of state for Richard Nixon. At the time of the hearing, Rogers was also the attorney for Lockheed Corporation. This was to be the last word in what actually happened to the Challenger on January 28, 1986. However, within all the testimony and paperwork, charts, documents processed by this commission, many things did come out, much to NASA's anguish. The Commission started June 6, and took months to complete.

What did we really get out of the Rogers Commission review of the Challenger Disaster? Not much information, at least not as much as we should have gotten, out of all this intensive review was seen. The Commission really failed to find out the root of the issues that caused the explosion. The Commission concluded that the concerns of Thiokol's engineers regarding the cold weather were not relayed to those who made the final decisions for launch. In short, it was a failure to communicate. However, that is just too easy. While some portion of the Rogers Commission were decidedly critical of NASA, the commission was sure to note that they had gone through every record that was available. The statement "… serious flaw in the decision making process leading up to launch" is not enough to substantiate what occurred. In fact, if anything was seriously flawed, it was the Roger's Commission itself. The commission released NASA officials, and that is the top tier of officials, of any direct responsibility for the disaster. That in itself is the biggest problem with the Rogers Commission logic. How is it possible to not lay the blame at the feet of the people that were in the lead in running the Shuttle program and Launch program? The last time that this author looked, the statement, "the buck stops here" (aka President Harry Truman), should sum up the responsibility of those in the front office at NASA. Did NASA's top tier of decision makers know something about how to exempt oneself from that responsibility? Perhaps the Rogers Commission didn't understand the nature of their job. The Rogers Commission did ignore substantial evidence most of it that was presented to the commission via private means and some at the public hearing itself. Those NASA officials were fully aware of the long history of problems that led to Challenger's demise.

The Commission totally left unchallenged many of the statements that NASA officials made that were not only contradictory but also directly confusing. At some point in the hearings, the commission allowed witnesses from NASA to be coached on how to deal with some of the tough public questioning that they would undergo. It wouldn't do to allow a NASA official to be examined at this public affair and tell the honest, unabashed truth, now would it? As far as the question of why the final decision to launch was made, the commission ignored so many suspicious coincidences and left so many questions unanswered that further investigation was needed, but never attained.

The commission did allow much of the launch and the internal problems of NASA to come out in the wash, however it wasn't anything earth shattering, as it should have been considering the issues involved. Much of the material was leaked to the press. The extent to which the Rogers Commission avoided drawing the obvious conclusions and asking the questions suggested by the evidence was noted. James Beggs, NASA Administrator of sorts, (remember, he was on leave due to some legal issues), said in an interview with Washington Monthly Magazine that the commission, *"looked at it from the bottom up, but not from top down."*

The White House has its say

President Reagan's choice to chair the Commission with William Rogers, Lockheed's attorney said, "We are not going to conduct this investigation in a manner which would be unfairly critical of NASA." Rogers continued with, "Because we think, I certainly think NASA has done an excellent job and I think American people do too." The selection of the other panel members was questionable. Even though the commission was created to investigate NASA, NASA chose the panel members. This seemed to be a little out of line, NASA investigating NASA. This happened once before with the Apollo 1 Fire, NASA investigated itself. Acting administrator William Graham, according to the Orlando Sentinel newspaper, said that the nominated commission members, most of who were accepted by President Reagan, are all familiar with spaceflight. Yet how objective could they be, with seven of the thirteen members having direct ties to NASA in the form of astronauts, former astronauts, NASA consultant, designer of the shuttle engine and former director of the Pentagon shuttle program along with executive of one of the companies that was a NASA sub-contractor. It was possible that this group would do an objective job of getting to the truth, regardless of how it would affect NASA. However, the chances of that happening were slim to none, at least in any form of reality. The true test of the success of this panel would be found in how it would investigate both the decision to launch and the history of the O-ring problems.

The Commission opens

February 6, 1986 found the commission gearing up for a start of the hearings. NASA's own internal investigation focused on O-rings as a probable cause of the accident according to the press at the time.

The first hearing began with NASA's top official not mentioning the O-rings as a probable cause of the disaster. Acting Administrator, William Graham said simply that" NASA continues to analyze the system design and data, and as we do, you can be certain that NASA will provide you with its complete and total cooperation." Along with that statement, associate administrator Jesse Moore, the man who was responsible for the final launch decision said that in their search for the cause of the accident "the status as of today {is} we have reviewed some data and our analysis does continue." Moore also said that before the launch, his only specific concern was that the low temperature might affect the water systems on the launch pad that were used for sound suppression, including the system that allowed technicians to wash out their eyes in case of accident.

The director of the Shuttle project office, Judson Lovingood, told the Commission that there had in the past been O-ring anomalies, but they had been "thoroughly worked." There had never been a case in which both primary and secondary O-rings had eroded. That was an untruth from the word go.

Arnold Aldrich, the National Space Transportation System Director from Houston and one of the top officials involved in the shuttle, testified that, "we had no concern for the performance or safety of the flight articles (Orbiters, boosters, main engines or fuel tanks) at the time, no do I even at this time." That was taking a huge step into fantasyland. Memos on the O-rings had been going back and forth, along with the Thiokol tests finding the charring and eroding reaching to the secondary O-ring. It was listed in almost every flight of the shuttle made.

A closed hearing of which the transcripts only recently became public, NASA officials began to discuss their knowledge of the history of the O-rings problems. Aldrich, the number two man on shuttle launch chain of command, told the Commission the O-ring erosion "has been in discussion in the program at least during the last year, and that was the worst case of erosion occurred in a cold weather launch a year earlier.

A *New York Times* story that was published February 9, 1986 was based on a memo written by Aldrich, seven months earlier, which described the history of the O-ring erosion. The memo cited engineers concerns at NASA headquarters that flight safety was compromised by potentially catastrophic O-ring in-flight erosion. The *Times* mentioned several other supporting documents including some documents on the analysis of the O-ring charring. Engineering reports stating that the back up O-ring couldn't be relied on, were in black and white. How could the top officials from NASA not be aware of these documents? The documents showed the O-ring problems were discussed at all levels of the agency. Therefore, it seems either some of these top tier officials were lying, or they were negligent in their reading.

These were just the first sheets of a long paper trail of evidence that NASA's top echelon were in serious denial of telling the truth about knowing of the O-ring situation. You would have expected the commission to point this out to NASA officials that not describing during the first part of the hearings the seriousness of the O-ring problems and not correction those problems before the next shuttle launch. Instead, it seems the Commission has coached the top line officials on how to avoid being embarrassed when this information became public. Why? NASA's top line officials knew, why were they being protected of exposure for their negligence.

February 10, 1986 brought about a closed hearing, after Jesse Moore (Associate Administrator of Space Flight), conceded the basic accuracy of the *New York Times* story was correct. William Rogers seemed to have counseled Moore on how to handle potentially tough questions about the why and wherefores of the agency not correcting the O-ring problems if they aware of them. Rogers said to Moore, "Now, everybody recognizes that you are going to make mistakes in judgment but at least you have to show that it wasn't done in a careless fashion and that there were meetings and you thought abut it and who was there and things of that bend." That was quite a statement, sort of like a defense attorney trying to keep his client off the witness stand. Rogers also suggested to Moore that he might want to answer all allegations that the agency was lax in fixing what it formally acknowledged (the O-ring) as a serious problem.

Way back in 1982, NASA reclassified the O-ring seals from *"Criticality 1R"*, which meant that there was a redundancy feature that would protect the shuttle from a failure in a main primary O-ring back down to *"Criticality 1"* which meant the back up safety feature could not be relied on to prevent the catastrophe that did eventually happen.

William Rogers went on to say "We want to be careful that NASA doesn't suggest by (SRB project director Larry Mulloy's answer) that nothing has changed (after the reclassification) that would be a devastating comment. I think the answer to that is "We're not sure yet, that is why we're studying". What that statement said was Mulloy knew that the O-ring classification was changed to one of no redundancy, which condemned the main O-ring to support of itself. It seemed to be fine with Mulloy and other NASA leadership. Yet they all knew the O-ring was in desperate straits. Thiokol told them, their own NASA engineers told them, yet they faced the American public and the families of the astronauts with shrugged shoulders and a well-rehearsed statement.

More Panel Members

As the Commission continued other, panel members began their questioning. Commission member Lt. General Donald Kutyna said to Mike Weeks, (Jesse Moore's deputy) "My problem is the *NY Times* kind of problem. Here it says that Cook (Richard Cook, a budget analyst for NASA who worked on the Shuttle Program) says' it's going to be catastrophic and here is another guy who says loss of mission, vehicle and crew, the formal description of what would happen in event of an O-ring failure. Somehow we've got to be able to explain in the open session tomorrow, why this is different from what you said (that the O-ring problems didn't constitute a serious safety of flight issue)." At another hearing, after being told the Rockwell engineers had opposed the launch, William Rogers said, "If Rockwell comes up in a public session and says "we advised NASA not to launch and they went ahead anyway, then we have a problem." Yes, you could say that. It seems that the Rogers commission was the biggest problem by trying to cover up for NASA so that NASA would not lose face. Rogers had another problem of his own. The *NY Times* article seemed to know a lot more than his commission did about the history of the O-ring. It was not sitting too well with Rogers. However, he was still using a tone of apology instead or prosecutorial attitude regarding the questioning of NASA officials. Rogers was very gently massaging NASA officials to be more forthcoming. "This is not an adversarial procedure." Rogers said. "This commission is not in any way adversarial and we hope that in future, as much as is humanly possible, when you think information had been developed that we should know about that you will volunteer to give us that information." Good luck with that, Mr. Commissioner. The press reports and testimony went on to point out that NASA' early knowledge of the O-ring problem was evident but not forth coming from the lips of NASA officials. The commission began to become less protective of NASA.

On February 11, 1986, the Commissioner Richard Feynman and eminent physicist, blew the roof off the entire NASA fraud with one little experiment that he conducted on the Commission floor. He dipped the O-ring material into a glass of iced water and showed it had no resiliency when removed. This very small but dramatic test showed how serious the O-ring problem really was, and how obvious it must have been to NASA officials.

The Commission moves on

Other commissioners on the panel began to question NASA management, attitudes and actions. One of those members was John Young, first astronaut to fly the shuttle.

On Saturday, February 15, 1986, William Rogers declared that the launch decision *"may have been flawed"*. He then ordered an internal investigative body set up by NASA to revise itself and not include anyone involved in any of the launch processes. It was a bit late for this action. From months of accumulating evidence, some of it uncovered by the commission, showed virtually the entire NASA bureaucracy *knew* about the O-ring issues. However, not everybody, including the Astronauts that flew the shuttle knew. The final report insisted that top-level officials who were responsible for launching Challenger were "unaware of the recent history of problems concerning the O-rings and the joint." That is hard to believe; however, that is what the Commission came up with. Evidence to the contrary was abundant. To start, there were statements by Aldrich and Lovingood in early hearings, the *NY Times* Article and memos. During February 11, 1986, in an open hearing which the report doesn't mention, Lawrence Mulloy told the commission that the April 1985 launch of the shuttle had caused erosion in the secondary O-ring, meaning that the primary O-ring failed totally. That was very heavy and frightening information to come out at this point in the investigation. To add to that, the O-ring charring was a major agenda item on all Jesse Moore's monthly staff reviews during 1985. How was it possible that Moore, knowing that this was going on, allowed Challenger to launch that day?

While at NASA headquarters in Washington D.C., engineers who worked for Moore, had been deeply involved in review of the O-ring problems during the 1984-85 period. It was one of those engineers who said in mid 1985 that they "held their breath" with each shuttle launch because of the O-ring, a statement passed on to the press and the commission. The report also didn't mention that they were told on March 28, 1985 that a top SRB engineer had been advised "not to list" O-ring charring on the Headquarters meetings as it was considered "too sensitive an issue" to put into writing.

It would be ludicrous to assume that Moore never knew the severity of the O-ring situation, or that he was oblivious to what his own engineers were doing.

More nonsense in the Rogers Commission

At an August 19, 1985, meeting at the NASA Headquarters on the subject of the O-ring, nothing is mentioned concerning the testimony in the July hearings of the House Science and Technology committee, (another Congressional oversight committee) of which Jesse Moore's deputy, Mike Weeks, chaired this meeting. He was noted as saying that Moore was told soon after about the situation on the O-rings. The commission concluded in its final report that the "O-ring Erosion" history presented at LEVEL I (Jesse Moore) at NASA Headquarters, in August of 1985, was detailed to require corrective action before the next shuttle flight. Why didn't that happen and how could Moore have such a short memory when asked by the Rogers Commission question on the O-rings. According to the earlier report, officials at NASA including LEVEL I officials, didn't know the history of the O-ring problem. When asked again by the Rogers commission on February 10th, about his knowledge of the O-ring problems, Jesse Moore admitted that he knew earlier flights had shown O-ring erosion, but he said he didn't this it was a safety of flight issue. Sally Ride, astronaut and panel member for the Rogers Commission asked, "What amount of erosion would have given you a problem to call it a safety of flight issue? She asked this question of Mike Weeks, deputy to Moore. Weeks replied:

"Sally, I don't think you should get the idea that we weren't deeply concerned about the first instance of the secondary O-ring erosion." Yet, the commission refers frequently to the Flight Readiness Reviews. It does not mention that it was normal NASA procedure for James Beggs (absent NASA administrator) and Arnold Aldrich (Shuttle program director from NASA Houston) both cleared by the commission of any prior knowledge of serious O-ring problems, to attend the FRR. These reviews regularly included discussion of the serious O-ring problems.

The commission ignored the role of the top NASA officials in making three critical decisions about the O-rings. In 1982, NASA upgraded the O-ring classification of seriousness from "Criticality 1R" to the highest degree of potential hazard "Criticality 1"

In the commission, report "LEVEL II" (re Aldrich) agreed to the decision of August 1985 in which NASA decided that even though it had acknowledged that an O-ring failure could be catastrophic, the shuttle could "fly as is". The Rogers Commission report did not mention under normal NASA procedure, William Lucas, the director of Marshall Spaceflight Center, who also was cleared of any knowledge of the O-ring problems would have played a critical role in making any such "fly as is" decision. The question begs to be asked: Who was responsible if all these top NASA officials had no idea of what was going on? The entire scenario is ludicrous and disgusting.

The knowledge of the O-ring problems was so widespread and considered so serious that NASA with top officials of the agency approval, had already begun testing an improved joint design.

The *NY Times* reported the agency ordered seventy-two new booster rocket casings to accommodate the new joint design that was being tested.

Questions anyone?

The second major failure of the Rogers Commission was not determining why the launch was allowed over the protest of Thiokol the SRB contractor. It concluded that those who gave the final approval for the launch did not know that Thiokol engineers had recommended the launch be halted for fear the cold weather would have a deleterious effect on the O-rings. There were many reasons for the contention that none of the upper level officials knew of Thiokol's objections and it is difficult to understand or believe. The NASA launch procedure calls for decisions and objections to methodically follow a prescribed chain of command up to the three levels of decision leaders.

Stan Reinhartz, another top official that wasn't exactly high on the list of engineering specialists was the Marshall Spaceflight Center Shuttle

Project director. He was a LEVEL III official. The Monday evening before the Tuesday launch of Challenger, Reinhartz claimed he visited a motel at the Merritt Island Holiday Inn, near the Kennedy Space Center. This was so he could tell his boss, the famed and feared William Lucas, that there was going to be a teleconference during which they would prepare to notify the LEVEL II official, Arnold Aldrich, of Thiokol's concerns. Lucas, as director of MSFC, reported directly to Jesse Moore. Larry Mulloy, the SRB chief and James Kingsbury, science and engineering director, were also involved in this chain of command. Reinhartz however, didn't inform Lucas of the intention to go to LEVEL II with Thiokol's O-ring objections.

He said he just told Lucas that Thiokol had expressed some concerns. That is really a good one! Lucas said that he merely asked to be kept informed. The care and concern of these gentlemen is really mind blowing.

Reinhartz said that he did not tell Arnold Aldridge or Moore about any of the issues because he saw Thiokol's expression of concern as a "routine matter" that was resolved. Can someone explain where it was resolved?

No, we didn't think so.

With not discussing these issues at the LEVEL II or I offices, Reinhartz had gone against instruction given by Moore earlier that afternoon to report constraints on the launch. If this was a routine problem, why did he visit the motel? Why did NASA require written approval from Thiokol headquarters in Utah, as unprecedented procedure? Why did NASA officials put such pressure on Thiokol to consent with one official articulating the memorable sentence, "My God, Thiokol, when do you want us to launch, April?"

Under these conditions, Reinhartz, who was a mid level manager, green as a granny smith apple when it came to space operations and new to the job, just didn't feel it was important enough to tell his bosses that there were some very serious issues on the table concerning the launch. It's hard to fathom that he would make such a decision on his own with just 5 to 6 hours before a launch that morning. Reinhartz and Lucas worked next to Aldrich and Moore. Yet, all parties claimed that not a word was spoken on the O-ring controversy that had taken some ten hours of discussion the previous afternoon and night.

Their description of events is hard to believe but it is also hard to disprove in part because the commission was so weak in questioning the witnesses. For instance, William Rogers asked Reinhartz about the discussion in the hotel room:

Reinhartz: "I discussed with them the nature of the teleconference, the nature of the concerns raised by Thiokol and the plans to gather the proper technical support the people at Marshall for examination of the data. I believe that was the essence of the discussion."

Rogers: "But you didn't recommend that the information be given to LEVEL II or I?

Reinhartz: "I don't recall that I raised that issue with Dr. Lucas, I told him what the plans were for proceeding. I don't recall Mr. Chairman making any statement regarding that."

Rogers was getting to a very urgent issue. When Reinhartz told his superiors about the war going on at Thiokol regarding the condition of the O-rings, Rogers just dropped the discussion. There was no follow up questioning like, "Did you tell Dr. Lucas that the Thiokol objections were serious enough that engineers had actually recommended not launching Challenger?" or "Did Lucas indicate that he wanted Thiokol to change its mind?"

Information was coming out, but it wasn't being followed up on to draw a deeper conclusion. Rogers had rebuked Lucas for hoodwinking him on what really went on with the teleconference. "Your describing the telecom as though it were just sort of one of those ordinary things and I don't believe that it is accurate." Rogers went on to ask William Lucas of Marshall why he didn't go to his bosses, Aldrich and Moore. Lucas said he thought the issue had been resolved and that he was not a part of the formal decision making process. Where did that come from?

Rogers again asked, "You had occasion though to talk to both Mr. Aldrich and Mr. Moore before the launch." Lucas answered, "Yes sir". Rogers continued with, "And whether it was in the line of authority or not, you had ample opportunity to pass on the information that there has been a serious concern about the seal, isn't that right?" Lucas answered: "Yes sir. I had the opportunity to talk to them." Rogers continued with, "Okay, I have no further questions."

Rogers once again, almost made a valid opening into the chain of command question and he shut it down before it could be continued.

The big question remains...WHY?

Conspiracy Theories

As if there wasn't enough wrong with this investigation, there was the ominous cloud of conspiracy theory hanging over the proceedings. There were so many contradictions and irregularities in the testimony that was given by the NASA top-level officials, it would be enough to raise serious doubt in anyone's mind what was really going on here. Did NASA's elite intentionally mislead and misconstrue the information it gave to the commission and the public.

There is other testimony that buttresses the possibility that there was some sort of cover up by NASA. Allan McDonald, who was a Thiokol engineer and who had strenuously objected to the launch for safety reasons testified that Larry Mulloy, the NASA official credited with aggressively pushing for the launch, had later tried to coerce McDonald. McDonald said, "Mulloy came into my office and slammed the door."

McDonald continued during the May 2 hearing, "and as far as I was concerned, it was very intimidating to me. He was obviously very disturbed and wanted to know what my motivation was—I won't use his exact words for doing what I was doing — (cooperating with the commission)." Mulloy said; "As I understand it, you're giving information to the commission without going through your own management, without going through NASA and what's your motivation for doing that?" McDonald told him, "And I told him to calm down, that I didn't think I had to get a note from my mother or anyone to give anybody information and I felt it was appropriate to give them information."

Another panel member Robert Hotz interjected: "Did you get the feeling that there might be some feeling on the part of Huntsville (MSFC) people that they wanted to control this flow of information to the commission?" McDonald said: "I got the feeling that was happening."

To add to this, a letter was written by a Marshall employee signed with a very theatrical "Apocalypse." This anonymous letter was read with a bit of salt though it seemed to have correctly predicted NASA's official conduct. The author of the letter really did show a tremendous knowledge of the internal workings of NASA/Marshall managerial methods and how booster rocket problems were solved. It gave a detailed description of a private meeting that the writer claimed to have been at with William Lucas, head of Marshall in attendance. At this meeting, the plans for a cover up were being laid out.

"Under Phase I of the cover up information was to [be] withheld as long as possible, then fed to the press piecemeal." This was according to the contents in the letter. It was rationalized that the longer the information could be withheld, the better, as the course of "world events" would ultimately weaken the initial shock and public reaction. It continues that one data could not longer be held back, Phase II would be to present as much of the highly technical data as possible. In essence, snowball the public and press with lots of tech talk. Again, this would cause the public and press to continue to kick material back and forth with various diverging arguments regarding the released material. The results would be confusion, misinformation and basic chaos.

The stories would be planted which would serve to pull all the blame away from Marshall and dump it on Thiokol, and the contractors doing the entire shuttle processing at KSC. However here we are again, the Rogers Commission lost the ball and only awkwardly chastised NASA for not being forthcoming with information. The statement claimed that, "for the first several days after the "accident"—possibly because of the trauma resulting from the "accident"—NASA appeared to be withholding information about the accident from the public."

Holes you could drive a truck through

The biggest holes in the Rogers commission report was not in its failure to explore if there was a cover up or some conspiracy but why there was so much pressure to launch the morning of January 28, 1986, given the conditions of ice and cold. As the former administrator James Beggs said; "The launch decision was made in the face of quite a lot of adverse conditions," Thiokol's serious concerns about launching giving what they knew about the O-rings and Rockwell's concerns about launching because the ice on the launch pad was at a dangerous level, just doesn't add up to NASA's decision to go ahead. According to what NASA thought, or let's say presented to think was that this was a regular launch technically. There wasn't anything going on that was of any concern. That's what they wanted everyone to believe, especially the press who had been hounding them for missed launches and delays. However, this was not a typical launch and that is what is so confusing. Every single item that the Rogers commission, the newspapers, the everyday, average person who took one look at that ice hanging off of the pad knew something was not right, so why did NASA just ignore all the signals? Alan McDonald of Thiokol, one of the several people interviewed by the Commission and involved in what Thiokol saw as a danger said, "I've been in many Flight Readiness Reviews, and I've had a very critical audience…justifying why our hardware was ready to fly…. I was surprised that the tone of the [pre-launch] meeting was just the opposite of that. I didn't have to prove I was ready to fly…. In this case, we had to prove it wasn't ready and that's a big difference." Yet in the case of Challenger, NASA, who was usually extremely cautious about decision making, while knowing there were problems with the O-rings via previous experience, just passed the buck, closed their eyes and lit the candle. On January 27, 1986, NASA hammered on Thiokol to agree to conditions that Thiokol was already telling NASA were deadly. There had never been a launch below 53 degrees and here was the shuttle on the launch pad for days in temperatures well below that. The day of the launch, the temperature was 28 degrees. Thiokol's vice president, Joe Kilminster in signing the approval document that the company agreed with NASA's decision to launch, signed Challenger's death warrant. That's not being dramatic, it's being truthful. Thiokol in the offline conference agreed amongst themselves launching was not in the cards. What pushed Kilminster to turn tail and sign? The bottom line monetarily for Thiokol comes to mind and the fact that who wanted to tick off a customer. However, was that all?

Pressures and more

We all know the story that NASA was under the gun for shuttle lateness, missed launches delays etc. Public relations were also pushing the teacher in space project. Did this project have anything to do with pushing Challenger off the pad? Was it the fact that the State of the Union address had been scheduled to include a statement regarding the Teacher in Space project that would have ended on the day of the address? There was widely reported speculation that a White House official, possibly the Chief of Staff, Donald Regan gave the orders to NASA to, "Tell them to get that thing up." However, the press secretary, Larry Speakes, very angrily denied the report and the evidence that this happened has never been turned up. However, even without a direct order, the timing for the State of the Union address did create enormous pressure within from the White House or inside NASA's official top tier of managers to do just that. NASA, since the dawn of the Mercury program had always been a Public Relations agency. Anything that could be done to make the agency or the program shine in glory, it was done. The Teacher in Space program was PR gold and it really would rally the public to support more funding for the programs NASA had on the back burner. It didn't hurt pushing Congress for support either, and pork barrel projects were always a Congressional favorite. At some point, NASA had become increasingly sensitive to how the media was portraying them. As of late, the press was giving them hell and making them look like a bloated, incompetent agency. After the glory days of Apollo, NASA was hoping the shuttle would be their meal ticket to improved things. In an interview with the Washington Post newspaper, after the Challenger disaster, KSC director Richard Smith pointed out that the pressure from the news media had a powerful influence on the atmosphere in which the launch decision was made. Because of all the delays with Challenger and the Teacher in Space program, it never made the State of the Union address. The one-day delay that went very back to December meant that the Challenger flight would have still been in orbit when the State of the Union address was given. NASA, to hedge their bets, had already submitted some sparkling material to the White House for use in the State of the Union address, *"Tonight while I am speaking to you, a young secondary school teacher from Concord New Hampshire is taking us all on the ultimate*

field trip, as she orbits the earth as the first citizen passenger on the shuttle. Christa McAuliffe's journey is a prelude to the journey of other Americans and our friends around the world who will be living and working together in a permanently manned space station in the mid 1990s, bringing a rich return of scientific technical and economic bene-fits to mankind. Mrs. McAuliffe's week in space is the first one of the achievements in space which we have planned for the coming year."
Hence, if there was "no" White House pressure to get this launch up on time, it seems a bit strange to have such an elaborate statement for a State of the Union address that wasn't going to matter. The White House, including Larry Speakes and Patrick Buchanan, the communications director who wrote the speech said the submission wasn't needed and was forgotten about. The State of the Union address contained nothing at all about the Teacher in Space program and that President Reagan was not going to mention McAuliffe while he stood before Congress. However, none of it ever happened because the Shuttle blew up 73 seconds after launch. However, were that pressure, non-existent according to some, and a large part of why Challenger was allowed to leave the pad still ques-tions the ice issue. Some other small details were not discussed in the press and barely addressed at all by the Rogers Commission. The stance cancellation of the Challenger launch the preceding Sunday. NASA again followed a procedure unprecedented in its history.

Due to the weather at KSC being unpredictable at that time of the year, astronauts would be put on the shuttle even if there were bad weather. The reason for that was the weather's changeability. It could be raining in the early morning but it could clear before the launch window closed that day. In this case, however, NASA officials canceled the flight that Saturday night because there was bad weather in the area. That meant they were not even going to *try* for a launch that Sunday. As it turned out, Sunday was a sunny and warm Florida day; so warm in fact, the O-rings were in good condition and would not have leaked. Why was the launch canceled that Saturday night *instead* of Sunday at launch time? The possible explanation might relate to a technical requirement that the Commission did not discuss in session or otherwise. It had to do with the loading and unloading of the hypergolic fuel and the extreme shifts in temperature in the lines of the fuel tank that puts a big strain on the insulation. NASA's requirement was that fuel may not be loaded or unloaded more than twice in a 48 hour period. If NASA had fueled up for Sunday and gotten the astronauts comfortable in the Orbiter then canceled, NASA could try again on Monday. If that failed, NASA wouldn't be able to try again until Wednesday. Canceling the launch on Saturday, *before* NASA had fueled up, NASA increased the chance that the shuttle would fly on Tuesday night when the President delivered the State of the Union Address. Interesting?

The chairman of the House Science and Technology Committee, Representative Don Fuqua, who was at KSC Saturday night said the desire to get the shuttle up before Wednesday was a factor in canceling Saturday night instead of Sunday morning. Fuqua said that another delay might affect the flight schedule of the GALILEO Jupiter probe, which was supposed to be flown on Challenger in May. However, the GALILEO explanation doesn't sound right since NASA officials testified that there was a two week buffer zone before the shuttle delays would affect Galileo's launch.

The Commission should have delved further into the State of the Union issue and the GALILEO issue. These were valid questions and yet they weren't touched. The decision to cancel on Saturday was so off the point that even James Beggs thought the Rogers Commission should have been asking why. Beggs said, "It was a bad decision" concerning the Saturday cancellation. 'The Commission ignored or did not wish to address why did the launch [Tuesday] and why didn't they launch when the weather was perfect on Sunday.

Another question comes to mind. One person that should have been called to testify and wasn't, was William Graham, the acting NASA Administrator. Graham got basically no attention because he was not involved in the normal decision process for flights and had been acting administrator, replacing Beggs for only a month or so. Graham, who moved on to become the Science Advisor for the White House, came to NASA initially to help supervise the agency's involvement in Reagan's "Star Wars" program. Beggs really didn't want Graham as administrator because he really knew nothing about space. It was possible that Graham would not have had an objection to Monday night's teleconference issues raised by Thiokol. If there was any political pressure from the White House, it most likely came from Graham. Representative Fuqua said in an interview that Sunday Jan 26th, Graham told him that he took direct responsibility for the unprecedented weather related Saturday night cancellation, which is highly unusual given his previous lack of involvement with any launch decisions at all! To add to that, Graham sent Representative Ed Markey records of his phone conversations before the Tuesday launch in which he excluded calls made that weekend. There was likely to have been some White House calls since Vice President Bush was supposed to be at the launch on Sunday. The phone records were released and did include two calls to the White House. One call was to Richard Davis, the NASA liaison and the other call to Alfred Kingon, the cabinet secretary. Yet, Davis and Kingon were never called to testify before the Rogers Commission.

The most interesting aspect of the Rogers Commission investigation into White House pressure was that they never interviewed anyone from the White House. The commission bypassed that to ask NASA officials whether they had felt any pressure from the White House at all.

In conclusion—Failure

The Rogers Commission for all its investigation really didn't tell us anything more than we already knew about the conditions at NASA.

If anything, there was more that the Rogers Commission didn't tell us or didn't bother to dig out. There were many chances to ask pertinent questions of witnesses and officials, but William Rogers and his panel didn't do their job. The entire Commission was summed up with Richard Feynman's test of the O-ring material in a glass of iced water put the lid on the entire issue. There was problem with the O-rings and that was is. Of course, it isn't quite that simple. The Rogers Commission, however, did let us down. It didn't put NASA on the hot seat to be responsible for not guaranteeing flight safety for the astronauts and the orbiter. NASA knew there was a long list of problems with the O-rings and chose to bypass a fix. Why? Again, much of it had to do with funding. There just weren't enough funds to go for a complete fix, so it was said. The Rogers Commission never resolved who was responsible for the decision to by pass Thiokol's objections to the launch. We know that Mulloy made the final launch decision, but truthfully who was behind him.

The Commission did not answer the question and didn't question NASA why officials were behaving so irrationally regarding this launch. From what we have seen, either the top tier of officials got the word from the White House and pressure to launch or they didn't. We have heard that the word was passed but even the head of NASA at the time, William Graham, wouldn't divulge his phone calls over that weekend until much later. The Rogers Commission didn't bother to delve into the question.

The Commission failed miserably to really find out what and why the Challenger launch was allowed to go with the circumstances that sur-rounded it. Erring on the side of caution would have been acceptable, waiting until the weather warmed would have been acceptable, yet the

Rogers Commission didn't bother to go any further in concluding anything more than the O-ring was susceptible to cold weather.

In what we have seen of the testimony, the events that occurred before and after the launch NASA was seriously at fault. We know that. However, the reasons and the names that should have been held responsible have not been divulged and the Rogers Commission made sure that the conundrum of who, what and where remained as is. The fact that the astronauts never really knew what they were facing, because they were never told the hazards of the O-rings, is unconscionable. That alone and in itself is equivocates to manslaughter. In a court of law, NASA would be held responsible. However, in the political world, this Commission did nothing but obfuscate the facts, deny the truth and cover up the perpetrators. It is not acceptable. The sad part about this entire scenario is that on February 1, 2003, it was repeated, and we lost the shuttle Columbia and her crew. Had the Rogers Commission not played into political hands, not withheld and refuse to explore the necessary information, perhaps something at NASA may have changed for the good.

It did not. On February 1, 2003 we lost the oldest shuttle Columbia and her crew.

Columbia the "Grey Lady"

The launch of Columbia, STS-107 was the 113th mission for the shuttle program. She launched under clear skies without much ado, except for a chunk of foam the size of a briefcase that slammed down into the left wing of the shuttle. What should be said is that the shuttle slammed into the foam and not the foam into the shuttle. On board the crew composed of: Commander, Rick D. Husband; Pilot, William C. McCool; Mission Specialist 1, David M. Brown; Mission Specialist 2, Kalpana Chawla; Flight Engineer and Payload Commander, Michael P. Anderson; Mission Specialist, Laurel B. Clark; Payload Specialist 1 of Israel, Ilan Ramon, were unaware of the foam hit. The foam had broken free from the external fuel tank bipod ramp. The camera tracking the liftoff caught the image of the strike, but it would not be until hours later that NASA would see the images. A similar strike occurred during the launch of Atlantis in October of 2002. On October 8, a few days after the launch back in the VAB (Vehicular Assembly Building) film lab engineers took the photos and passed them out onto a table to discuss what had happened. The images showed where the foam had hit on Atlantis, it landed on a metal ring covered with insulation that attached the base of the left side of the booster to the fuel tank. This is what happened again on Columbia. One of the NASA film reviewers, who was also on the Ice, Debris and Final Inspection team by the name of Armando Oliu, said, "So, all we saw was the debris coming down the external tank and then impacting." On that booster, near the attachment ring was a very critical electronics box that sent computer commands from Atlantis. This box transmitted a signal that swiveled the booster nozzle to steer the shuttle during the first stage of the flight and dumped the booster after it used all the fuel up. If the box were disabled by the impact the result would have been a major disaster. However, the booster worked well, the explosive bolts fired two minutes and then after launch dropped into the Atlantic ocean below. Atlantis was safe in orbit, but the foam strike just missed the significant little electronics box. In this case, there was no damage, so there was no meeting to discuss the foam hit. Armando Oliu continued with "We knew that there was no problem with the booster's performance, and we know that the debris did not strike the orbiter." However, another of Oliu's coworkers wasn't ready to just drop the issue.

Bob Page had seen Challenger die before his eyes. He was known for his attention to every detail and left a job at McDonnell Douglas to come to NASA and help the shuttles get back to speed. He looked at the film analysis, since he was the boss of the *Intercenter Photo Working Group*, an agency responsible for covering the filming of shuttle mission takeoff and landing. What he saw sent chills of Challenger's demise flooding through his mind. Page knew that what he was seeing meant some serious trouble. Page remembered, "If we had lost the electronics to the booster during ascent, it would have been a bad day."

Why we are bringing up the Atlantis mission is critical to the explanation of why Columbia died. The problem of foam debris is one that has been going on since the first shuttle flight. It wasn't anything new, but the severity of it was becoming a large issue that NASA again, refused to acknowledge much like the O-ring situation. Out of seventy-seven shuttle launches, when photography was used, sixty-three flights had foam loss. The rest of the flights were night launches or cameras failures. What could have occurred was the foam hit in areas that the cameras didn't cover. There were some golf ball size dings found in the 24,000 protective tiles that covered the Orbiter. It didn't take much to damage a tile since the silicate material it was made of was brittle.

Early Foam Problems

In April 1981, in Columbia's first flight, there was a cascade of foam that hit some three hundred of the tiles. In some twenty -one flights that returned and were examined after each mission, some one hundred and forty tiles were damaged and more than thirty of them have deep pock-marks that measured equal to an inch or so. When balanced out, there had been some one hundred hits on the Orbiter's belly, including some of the under surfaces where the very crucial black tiles were laid. These tiles were exposed to temperatures of 2300 degrees. The black tiles measured six inches square and between one to three inches thick. Many times the pockmarks would pierce down to the base of the black tiles, which was getting uncomfortably close to the Orbiter aluminum infrastructure. Another question while the discussion is open, returns to why the Orbiter was not made out of titanium. The NASA response to that was:

• "Build it like any other aluminum airplane"

• Most of NASA didn't have the knowledge to deal with the many variants of titanium alloys available.

• Titanium is only as good as the alloy used to reflect whatever purpose it was being made for. While at the time of the shuttle construction, the A-12 Blackbird, the SR-71 Blackbird, the YF-12 Blackbird, the B-58 Hustler, the XB-70 Valkyrie were all in production and flying. Was it a question of money and lack of education that decided the Orbiter should be built of aluminum? It was most assuredly a mix of the two. Should the shuttle infrastructure been constructed of Titanium alloy, it would have had a better first line of defense against the tremendous heat of reentry.

The Orbiter tiles, which were made from reinforced carbon-carbon, protected the leading edges of the Orbiter nose and wings hence allowing those spots a better chance of surviving a direct hit by foam. In 110 missions, the Orbiter was subject to foam strike in each case. The October 7 launch of Atlantis had foam strikes that were endemic to shuttle liftoffs from day one. Why was no one at NASA paying any attention? **In NASA's Volume 10 Book 1 of the National Space Transportation System 07700 Document, which is also called the "Shuttle Bible", listed ground and flight requirements:** *"The Space Shuttle system including the ground system, shall be designed to preclude the shedding of ice and or other debris from the shuttle elements during prelaunch and flight operations that would jeopardize the flight crew, vehicle, mission success or would adversely impact turnaround operations."*

Apparently, rules were never enforced. In 1988, shuttle officials determined that foam and other materials falling off from the external tank and hitting the Orbiter *"WAS AN ACCEPTABLE RISK"*. That was an outrageous statement. It was documented that there was a *"remote chance falling debris from the External tank could cause catastrophic damage."* The statement went on: *"Potential effects noted in the report included damage to the Orbiter's thermal protections system structural overheating and loss of vehicle, crew and mission."* According to the documentation, NASA was literally playing handball with a live grenade. While the shuttle was built to sustain launch and orbit, getting it off the pad was the real danger. However, the foam was necessary and it was not an issue of dumping it. It provided the external tank with a necessary outer layer of thermal protection, which was almost two inches thick in some areas, to keep the propellants from boiling off. The temperature of the propellant was 423 degrees below zero. The insulation also helped to keep ice from forming. However, with the humid Florida weather and sometimes-cold Florida weather, as seen in the Challenger disaster, it was always an issue. There were four different types of foam While they all had the same polyurethane base, it looked and felt like the poly styrene used in cheap coolers and packing material.

How foam was applied

The Michoud Assembly building inn Louisiana was responsible for coating the External Tank, as it was responsible for building it. A computer system designed to control the coating of the tank, used spray guns. The foam was applied by machine and human power. When the tank arrived by barge at Kennedy Space Center, additional foam was applied by hand and spray gun to the areas that were cracked in transit or fell off. On November 19, 1997 which was Columbia's first launch, foam dropped from two areas on the ribbed mid section, causing some three hundred scratches in the foam. This incident was called "popcorning" and was a trait that was ascribed to the removal of freon gas from the process to make the foam more environmentally friendly. Let us stop for a moment and examine that statement. The removal of Freon gas significantly changed the makeup of the foam making it more susceptible for scratches and breaking off.

How many shuttle launches were there in a year, possibly three maybe four? Why would you jeopardize the crew and the Orbiter on a product that was used for no more than 2 minutes in launch and sat on the pad, untouched for a month at most and fueled only twenty-four hours before a launch. What was the environmental impact? It cost NASA much more money to produce the "environmentally friendly" foam than it did the regular foam.

What environmental affect did the Freon gas really have on anything? There are thousands of air conditioners and air-conditioning units, refrigerators of old designs that are still being used today. What affect do they have? There are more dangerous situations caused by this and car emitting fumes for not being tuned properly than could ever be caused by the launch of a shuttle. Industrial waste is being tossed into rivers every day in this country without anyone stopping the companies doing it, yet NASA felt it necessary to change the minimally least offender and jeopardize the safety of a billion dollar Orbiter and crew to kowtow to a group of "environmental greenies". It was a stupid, costly, politically correct decision on the part of NASA and in the end, it cost lives. The problem of removing the freon gas made engineers change the thickness of the foam in places where it either fell off or cracked. They made holes to allow any trapped gases to escape and sanded of the crust formed on the outside on the foam.

While all this was going on, NASA saw fit to fly ten more missions to be exact and at that point, NASA deemed shedding foam was "acceptable" again. What was it acceptable? Was it just an endemic part of NASA's political correctness? Why was nothing ever said about it?

Bipod ramps

There was a problem area on the external tank and Orbiter truss that had never been resolved causing many issues. That area was the bipod ramp that attached the Orbiter to the external tank at midsection. Two metal spindles were covered with foam. These spindles were attached to the tank just below the Orbiter's nose. They had a purpose: the struts provided surface that protected the spindles and hardware from the aerodynamic forces that occurred at liftoff and protected the spot from heating during the push to orbit. It also prevented ice from forming before liftoff. The ramps were created by spraying layers of foam on the spindles then trimming to shape and size approximate. Regardless of how hard workers tried, voids in the foam were sometimes left between the ramps and the coating that covered the tank's aluminum alloy skin.

While there was no documented case that foam fell off that bipod ramp on the right side of the tank, the engineers felt that a 17" wide liquid oxygen line helped to shield the right ramp from the launch stress. The left ramp shed debris about three times at least that the shuttle engineers knew about, before the Atlantis mission, which had some of the biggest chunks of materiel to fall off. A 9"x12" piece of foam on the left ramp came off in a 1983 Challenger launch. While engineers never figured out why, a few missions later, the ramp design went from a 45-degree angle to a 30-degree angle. In January of 1990, a large piece of foam including part of the left ramp fell off during the launch of Columbia. Engineers said the shedding was caused by air trapped in the voids between the foam and materiel under it. The cure was to poke small holes in the foam around the bipod area.

June 1992 found a huge piece of foam, 26"x10", fell of the left ramp during launch of Columbia and hit her. After the mission completion and landing, the area damaged was the biggest ever seen on the shuttle tiles. Shuttle engineers blamed the poor venting and drilled more holes to compensate. The foam strikes seemed to get larger year by year and they were just accepted as an everyday occurrence.

The engineers felt that the Orbiter would survive, and the external tank engineers felt confident they knew what was happening and confident that they were doing enough. Hence, no one really gave the whole "loss of vehicle" scenario a second thought even though it was written about in Report 37. While the Orbiter, managers were living with the thought it was possible to seriously damage the tiles fatally, and the RCC surfaces with huge foam hits from the bipod ramp were all reasonably viable. NASA didn't do much. There were only three cases on the books concerning the foam hits from the left ramp. There was no proof that there were any from the right side. There were studies that showed it possible that a ramp could break off and hit a critical part of the Orbiter.

However, the report felt that the possibility was negligible because of the air flow around the Orbiter during lift of and flight. Should large chunks of the ramp fall off, the foam was light like styrofoam. This of course, was not taking into the equation that the shuttle was traveling and the foam was speeding up, hence when it hit the Orbiter, it did it with enormous velocity. The ideas resolved from the first shuttle launch went on to make NASA feel that the foam hits were nothing more than a maintenance issue that made for considerable paperwork. It wasn't critical when it should have been considered "Criticality 1". One of the shuttle managers said, "We had this mindset that this was a nuisance maintenance. It wasn't a flight safety issue." These were famous last words.

Changes

It wasn't until after the photos taken by the astronauts aboard Atlantis showed the partially missing ramp did the External tank managers at Marshall Space Flight Center start thinking about making some changes. One of the plans had to do with removal of the silicon based coating under the bipod area that covered the skin of the metal tank, before the foam was applied. It was called "super light weight ablator or SLA (sounds like slaw). It had the consistency of chocolate syrup when applied and hardened to the consistency of cork. The SLA was added to areas for protection against high heating in the early days, yet later studies showed that it wasn't really needed in many locations. The SLA had already been removed from several propellant lines and other parts of the tank. The main worry in the bipod ramp remained in the air pockets. More voids formed between the SLA and the foam. When the fuel tank was filled for launch, the air in the voids was chilled so low that it became liquefied. The fuel levels dropped and the outside of the tank heated up after liftoff. The liquid expanded and became a gas again, which created pressure and popped off pieces of the foam. The SLA was a failure and the engineers decided to can it.

The Tank engineers also looked at the process used to make the ramps. Under the old procedure, the Michoud plant used a spray gun to manually apply the ramp foam in a single eight-hour shift. They were also concerned that the procedure was responsible for the voids as layers of rapidly rising foam folded over each other. There was no test for voids. Many engineers proposed spraying the ramp more slowly, over several hours or shifts with the goal of getting more consistent results in applying the foam. Engineer Neil Otte said, "There was still a lot of discussion about exactly how do we go about improving the foaming process." *NASA's Program Requirements Control Board* called a meeting on October 24, to review an amount of technical issues including the Atlantis foam hit. The board was called to decide whether there were any critical failures during the Atlantis mission worthy of being called in-flight anomalies. The Intercenter Photo Working Group said the foam loss on Atlantis should be called an IFA but the Marshall engineers thought the designation was not warranted, claiming it was a "Freak event." On that last statement, foam had been falling since the first shuttle flight. How could it be considered a freak event? Was it because this was the first "recorded" time that it hit the bipod assembly? There were three previous bipod foam incidents as well as several others "out of family" debris strikes, and they were all labeled IFAs (In flight anomaly). The case needed to be made that these strikes were a dangerous threat to the orbiter, so why wasn't it? There were other IFA issues on the plate. One had to do with the maneuvering thrust rocket that malfunctioned in orbit, and problems at liftoff that had to do with the detonating explosive bolts that hold the boosters to the launch pad. As the items were discussed at various meetings, the case couldn't be made clearer that the foam strike on Atlantis was indeed extremely dangerous.

NASA went back to the processing records for the external tank used on the Atlantis mission searching for any design changes that might have caused the problem. There were no changes. Jim Halsell, Astronaut and man in charge of the KSC preparation of Endeavor, the next mission after Atlantis, said: "Do we need to look at anything on the pad right now?" Neil Otte, a KSC engineer said, "Jim, I guess we would answer that no. There is really nothing we can do to that bipod that would give us any more confidence that what we've got right now." That was a frightening statement when you really looked at it. The question of turning the issue of the foam debris hits into an IFA meant that would make it a critical issue that could cause a problem to the crew, the orbiter or mission success. Lambert Austin, a NASA System Integration manager at Johnson Space Center in Houston, felt the foam incident didn't meet the formal requirements for an IFA but it should be looked at. On and on it went, all around the mulberry bush. No one wanted to take the cause in hand and make a serious decision that maybe, just maybe there was something seriously wrong with the foam strikes and they weren't just a thing that happened.

NASA's Stars

Ron Ditttenmore was one of NASA's rising stars. He was an engineer and manager who knew the day-to-day operations first hand of the shuttle. Many held Dittenmore up as the talented manager, but many also felt that he was arrogant and condescending of those that didn't measure up to his "high" standards. In short, he felt he was better than anyone was when it came to the shuttle. In 1992, he left a job as flight controller in Houston and after working for a short time on the ISS (International Space Station), he moved on to the Shuttle program office. Many senior managers were impressed with Dittenmore and watched his quick climb to the shuttle program office. Dittenmore became the integration manager responsible for the overall coordination of flight payloads and operations. As part of his job, he also headed the Mission Management Team that oversaw shuttle operations during flight. He managed Shuttle Engineering office from 1997-99 when he took over as shuttle program manager. Some of his predecessors accused him of having "Go Fever" and rushing to launch. Dittenmore grounded the program two times to deal with a wiring and tiny fuel line cracks problem. He postponed the December launch of Discovery to make sure those twenty-year-old manufacturing records were all right even though no evidence of problems was found. Dittenmore wasn't convinced the foam strike met the criteria of an IFA. However, he directed the External tank project office to do a presentation of the problem at an FRR (Flight Readiness Review) meeting for Endeavor's November launch. Dittenmore said: [19] *"I wanted this briefed at the FRR so that the entire agency had a chance to listen to rationale and if they thought it was weak then they would raise their hands and tell us we needed to do more work. If the safety organization didn't agree with this, they had the opportunity to raise their hand and say we needed to do more work. That's what an FRR is supposed to do."*

Dittenmore checked the room and the teleconference included representatives from the safety office, Astronauts Office, engineering integration, space and life science office, contractors and the shuttle project office. Dittenmore said: "Does anybody else have anything they would like to ask or have done?" The only sound in the room was crickets chirping, which is a metaphor for silence. That was strange considering that there were engineers there who had other thoughts about the foam debris strikes. Instead of being declared an IFA, the ramp foam problem was put under "Plan/Studies". That wasn't saying much. Only the shuttle

---

[19] Comm check- The Final Flight of the Shuttle Columbia Pg 65 Michael Cabbage/Wm Harwood 2000 Simon Schuster

film review office was asked for recommendations four weeks after Endeavor's November 10th launch. Many of the engineers were frustrated but at least the FRR would see it again. That afternoon, Marshall Space flight Center managers were to review Endeavor and see if the External tank was ready to fly. There were project officials and Marshall groups. According to all the charts and one that was in error which said there were only two previous instances of bipod foam ramp loss, which left out Challenger's 1983 flight, and notes reviews of manufacturing and processing records showed absolutely nothing. Another chart showed analyses of how air currents could transport debris coming off the ramp on the left side and showed that foam from this area would not impact the orbiter. Then there was a final chart titled, "Rationale to Fly" There was no safety of flight concern at all. This was truly the personification of NASA's internal ability to completely ignore a problem while having it right out in the open.

Not everyone at the meeting was buying that, however. At the end of the meeting managers passed around a sign off sheet where engineers and officials certify that the external tank was ready for launch. When it came to the structure and dynamics division to sign off, the man in charge of that division, Peter Rodriquez, refused. Rodriquez questioned the foam impact including several strikes on Atlantis's left wing, of which we wanted an answer. "How do you know that the foam won't come off again?" "How do we know what the Orbiter's limits are?" Another safety manager, Angela Walker also refused to sign until Rodriquez was answered satisfactorily. The next day, Neil Otte, of the film department met with Rodriquez. He gave him a blow-by-blow account of the history of the impacts. He also told him how shuttle management never put any constraints to launch against a foam strike. Rodriquez finally signed the FRR statement along with the safety officer.

On October 31, the shuttle managers from all facilities met at Kennedy Space Center for an FRR on the Endeavor launch in November. This meeting goes on, before each mission, in the large conference room on the ground floor of the KSC operations & Checkout building. FRRs are run to make sure that the launch is flight ready and all factors have been dealt with as far as safety goes. All of NASA's senior officials attend. Before the Challenger mission, managers argued via teleconference about the Cape's freezing temperatures and O-ring problems. After the Challenger disaster, many felt the decision to see each other face to face instead of over the phone was a better plan. Every launch since 1986 was now face to face.

Bill Readdy, the top manager for all manned missions, led the November Endeavor launch FRR. Several technical issues on the agenda had some serious concerns. There were the fuel line cracks in the shuttles main propulsion system that kept the entire fleet grounded most of the summer. Engineers were trying to figure out a potential catastrophic electrical problem that fires the explosive bolts to free the boosters from the launch pad. The foam issue wasn't an issue at all, at least not at this FRR. Readdy started the FRR meeting that Thursday morning by talking to the group about the need for safety and paying attention to detail. Mission operations had their say. The External Tank group talked about the foam loss and other tank concerns. The head of the External Tank program, Jerry Smelser a NASA old timer of many campaigns, didn't seem too concerned about the foam hit on the Atlantis tank. This same hit had the film team at KSC choking on their coffee. Smelser said[20]: *"My reaction was that it was similar to Styrofoam lid of a cooler hitting the windshield of a car or truck. It was detracting but not dangerous. We*

---

[20] Pg 69 Comm check—The Final Flight of Shuttle Columbia pg 69- M. Cabbage/Mn Harwood 2000 Simon Schuster

looked at the density of the foam and it was very light. A piece of foam this size does not weigh much, although I don't have the background or engineering expertise or all the tools to predict what is going to happen with the foam once it comes off. However, from a practical engineer's standpoint, it did not appear to me than anything that light could do damage to an orbiter. That was a practical farm boy engineer's judgment, but that also was substantiated by the people who did the analysis."

Smelser went on in his presentation to state: *"There are two other known instances where we have lost similar amounts of foam from similar areas. Evidence is that was not a substantial contributor to the underside damage to the Orbiters."*

Smelser showed another slide in his presentation, which showed the bipod ramp before and after the foam strike. Printed at the bottom of the slide *"The ET is safe to fly with no new concerns."* These are famous last words. Smelser's presentation went on to discuss more about the foam shedding and the shuttle. For all intent and purposes, he was firmly convinced there was no reason to fear for the orbiter. *"Nothing about this tank we are about to fly makes it less probable or more probable of having this happen than the previous one. We've had three out of one hundred twelve that we know about. Therefore, we are in fact; as far as this phenomenon is concerned, we feel it's safe from a safety of flight standpoint. We can't guarantee that the foam won't come off."* Dittenmore, who was at the meeting, remembered the reaction of the shuttle managers. It was not much different from the response in the control board meeting a week earlier. Dittenmore said[21]: *"I don't think we were*

---

[21] Comm check- The Final flight of Shuttle Columbia pg 70, MCabbage/Wm Harwood Simon Schuster/2000

impressed with the presentation as being a thorough discussion of the problem."

Smelser finished and was questioned by a former shuttle commander and now head of the Safety and Mission Assurance office in the Washington headquarters, Bryan O'Conner. Bryan O'Conner asked Smelser if the foam size was too small to cause any damage. Smelser said; *"It was a big hunk of foam, but there was no evidence from looking at the Orbiter that this does in fact get over and become a safety of flight issue."*

O'Connor wasn't satisfied. He asked Smelser if the foam could be dismissed as a safety of flight issue. O'Conner recalled the 'Integrated Hazard Report 37' and asked if in fact it was a safety issue but one that had been accepted regardless as a potentially catastrophe risk. Smelser wrongly stated that the report applied to "ice debris". O'Conner asked for the hazard report and they decided to forward the discussion to another time. O'Conner stated: *"I didn't want to get into a semantics issue with him (Smelser), but I thought it was sort of an understated thing to just say that because to me, this whole topic of stuff coming off the tanks and Solid rocket boosters and hitting the orbiters was an ongoing catastrophic hazard in the program that hadn't been closed out. We haven't made that hazard go away."* After O'Conner got a copy of the hazard report, NASA's Safety chief went back to the topic. While O'Conner rushed to avoid any semantics, it did get to the point. Smelser agreed he should have said "accepted risk to fly instead of non-safety of flight issue". Yet, Smelser was still holding onto his original statement and that he was right. He said: *"Our review of the integrated hazard report and the External Tank hazard seems to be consistent with what I said with the exception of the fact that I used the wrong words."*

The debate was over, and Bill Readdy felt that the integrated hazard report be updated since it hadn't been revised recently and didn't address the foam on the ramp issue. Dittenmore went on to say that he felt the safety system review panel should look at the report to see if technical changes were needed. *"I'm not sure it needs updating but it sure appears that we need to review it again to make sure its accurate in its language given what we've discussed today."* After the entire going back and forth, no formal action was taken regarding to the foam. It looked like what happened to Atlantis wasn't causing any issues for Endeavor's November flight. However, much as foam strikes never held up a launch, NASA being NASA made sure the paperwork was up to date. The basic nature of foam debris hits would be noted, every "i" dotted, and "t" crossed.

Endeavor finally launched

The November 11 launch for Endeavor was delayed during the final countdown by a leaky oxygen hose in the shuttle's mid section. While trying to find the problem, workers installing an access platform on Endeavor's cargo bay, slammed the shuttle's robot arm, ripping part of its protective thermal insulation. Engineers examined the arm for damage while Ron Dittenmore held the flight up for more than a week to be sure the oxygen leak was not a generic fatigue problem affecting hoses throughout all the shuttles. It seemed like Dittenmore was looking for trouble everywhere except at the place that it had been pointed out to him. On November 23, 2002, at 7:50 PM, Endeavor lifted off into a night sky. Trying to see any debris falling was impossible. After Endeavor's landing with a successful mission under her belt, the checks on the bipod ramp look all right.

Marshall Spaceflight Center was having some issues of their own. Smelser's team had fallen behind on attempting to report fixes to the ramp problem. They asked Dittenmore for an extension and got one until the first week of February. This was a week after Columbia was supposed to land, completing her flight.

Dittenmore remembered[22]:

"They came in and said Look Ron, we don't have anything to really tell you on December 5th. Why don't you give us another six weeks or something like that? I said okay, I want you to be right rather than wrong. I want to have a meeting of substance rather than you just giving me a bunch of viewgraphs. So, I said I would give them another six weeks. Did I think (the foam debris) was an FRR question? No."

On January 9, senior managers met again for the Columbia FRR. The foam strike on Atlantis was already a forgotten issue. It never came up in the discussions. However, Endeavor's FRR still bothered Dittenmore. *"I look back on it now and say that was one of the things we really missed as an agency. We had a presentation that was not thoroughly discussing the risk. In addition, four center directors. Program management, contractors and NASA associate administrator for spaceflight and they associate administrator for safety for the agency said it was okay. Look-ing back on it, that was a critical juncture, where we as an agency—at the highest levels of our headquarters management and our program management, our contractor management, and our team sitting in that room—had a problem that was not thoroughly discussed. And we pressed forward"*

Columbia comes home for the last time.

February 1, 2003 was a Saturday morning. There were many that were up and waiting for the landing and the end of the successful mission of Columbia. However, it wasn't to be, as we now know. The foam debris, which had struck her on lift off so many days ago, was now in play. The foam that hit the wing was destroying Columbia as she tried to

---

[22] Comm check-The Final flight of the Shuttle Columbia—pg72, M Cabbage/Wm Harwood 2000 Simon Schuster

make it home. She didn't. She broke apart on reentry over Texas and disintegrated taking the lives of her seven crewmembers with her. Could this all have been avoided?

As we have seen in all of the previous material, the signs and portents of disaster were there to be seen. However, NASA again decided to disregard those signs, much as it did with Challenger and allowed the foam debris issue, which had been a problem since the first shuttle launch, to continue without a reasonable attempt to fix it.

We will now examine what the CAIB or Columbia Accident Investigation Board found. First, let's look at the email that was sent to the Columbia crew advising them of the status of the foam hit:

"Experts have reviewed the high speed photography and there is no concern for the RCC or tile damage. We have seen this same phenomenon on several other flights and there is absolutely no concern for entry."

The email was obviously incorrect. Again, the astronauts were not given the conditions of their spacecraft, as they should have been. Let us track backwards and look at exactly what did happen to Columbia on the morning of the launch

The Columbia Launch

At the Columbia FRR meeting that was held on January 9, 2003, there was no mention of foam debris or hits. There was a discussion about cracked bearings in the fuel lines. However, Columbia got tentative approval for clearance for January 16 with engineers working to complete all the final tests. There were some very strange occurrences marking the pre-launch of Columbia. Due to some bad, weather in Florida, the Beech Bonanza aircraft that some of the astronauts were flying from Houston to KSC in, slammed into the runway on landing. No one was hurt, but it was a scary situation. The next FRR went on and there were no constraining issues holding back the launch on the next day. Ron Dittenmore, put his blessings on the launch after tests showed that if any of the sixteen bearings in the shuttle's propulsion system were cracked, the chances of a large piece of metal breaking off and damaging an engine were minimal at best. A last minute concern about the structural strength of the attachment rings that help to hold the bottom of the shuttle's boosters to the external fuel tank were up earlier in the day at Johnson Space Center. The concern was not about the Atlantis strike, two flights earlier but the materiel used in the aft attachment ring was not quite strong enough as NASA wanted. Should a ring fail in flight, the base of a booster could pull free of the tank in repeat of the Challenger explosion.

Columbia's tanking meeting started and engineers presented a last minute analysis that showed it was safe for flight. The management signed off and the launch team was told to go ahead. The three-hour process to start tanking started at 3 :00AM on Thursday, the 16th about an hour behind schedule because of some minor problems. The engineers began loading the upper section of the giant external tank with 143,000 gallons of liquid oxygen, while the lower section was filled with 385,000 gallons of liquid hydrogen. By 5:30 AM, fueling was complete. The Ice Debris and Final Inspection team began their tour to inspect for problems on the launch pad and the shuttle stack, like ice that could break off during launch and hit the orbiter. The team did report a slight buildup of frost on the external tank's left bipod ramp but that wasn't unusual and no other problems were noted.

The astronauts were loaded into the orbiter and the countdown went on. At T-31 seconds, Columbia's four flight computers took over the last moments of the countdown. The external tank began to pressurize, and the shuttle's hydraulic system was running. There were no reported problems. Columbia was ready to go. There was some four million pounds of thrust just waiting to bust loose and head for space. Liftoff was stunning to say the least. The astronauts reported seeing a sudden flash of orange fire and smoke over the cockpit windows as the small vernier rocket motors lit to push the spent SRBs out of the way. This was a very complex mission with lots of work for all the astronauts. Things were going great in space but the nightmare was unfolding down at Kennedy Space Center.

Photographs

Back down at Kennedy Space Center, the rumor mill was running wide. There was word going around that Columbia had taken bad foam hit on lift-off. Of course, the ever-present press corp. started to salvo NASA Public affairs about the leaking story. Public Affairs had tried to keep things quiet and solid but they weren't having a very good time of it at all. They had their line of defense set up and were telling all that asked, "They assessed the strike and concluded it was not a threat." It would have been nice if they let their engineers, who were sweating, know that all was well because Public Affairs was busy living in "The Emerald City" and the Wizard had proclaimed all was well. The astronauts, much like Challenger crew, knew nothing of the situation and were not told, at least not yet. It took one week into the mission before there was any talk of the hit. At that point the powers that were decided they had better tell the astronauts something because they had interviews with the press coming up and it wouldn't look good if the astronauts were taken by surprise. The email read: [23] *"You guys are doing a fantastic job staying on the time line and accomplishing great science. Keep up the good work and let us know if there is anything we could do better... There is one item that I would like to make you aware of before the planned interview. This item is not even worth mentioning other than*

---

[23] Comm check— Final Flight of the Shuttle Columbia"Michael Cabbage/Wm Harwood pg 91

wanting to make sure that you are not surprised by it in a question from a reporter. During ascent at approximately 80 seconds photo analysis shows, that some debris from the areas of the (left side external tank) Bipod attach point came loose and subsequently impacted the orbiter left wing.... creating a shower of smaller particles. The impact appears to be totally on the lower surface and no particles were seen to traverse over the upper surface of the wing. Experts have reviewed the high speed photography and there is no concern for RCC or tile damage. We have seen this same phenomenon on several other flights and there is absolutely no concern for entry. This is all for now. It's a pleasure working with you every day."

A cold chill was working its way through Kennedy Space Center as the first of the photos started coming through. Marshall Space Flight Center was having its shock attack, too. A Marshall engineer called in at around 10 AM on Friday, after seeing the filmed images of the launch. It was a low a resolution television video from a U.S. Air Force tracking camera, which was placed some 26 miles south of the launch port. It had picked up the debris hit. The KSC film team had just loaded a sharper 35mm video from the same camera location that got it just about 1/2 earlier but that camera was hopelessly out of focus. Another high-resolution camera some 17 miles at the Canaveral Air Station was also examined. That is when the KSC engineers went into deep, unforgiving, shock. Exactly what was described in the email to the astronauts is what the engineers found on the videos. With all this information, it was still impossible to tell the severity of the damage. The engineers were falling over themselves trying to find an answer without much success. The Atlantis strike, only three months before made Bob Page try to find out more. He knew about satellites and telescopes that the military used. They could be trained on the shuttle to detect exactly what did happen. While these photos were classified, he knew that the photos had been taken they saw a panel fall off discovery in 1988 launch. Page was looking to beg borrow or steal to get what he felt he needed and those were the classified shots of Columbia. He knew that his colleague Wayne Hale. Hale was the KSC launch integration manager, which made him responsible for all launch preparations and the final go for lift off. Page asked again could get those shots. He had top-secret clearance and knew what buttons to push to get those images.

Page said, "This was absolutely a formal request of these pictures. We could not tell from the film that that we had if there was or was not damage to the vehicle. The resolution was not there." Page did ask Hale if it was possible to get shots of Columbia in orbit. Hale looked at him in silence. Hale just refused to talk about the classified aspect of the photos. Page asked Hale another way "We can't tell whether or not there was damage to Columbia and we need pictures of the vehicle in orbit." Finally, Hale agreed and said: *"I hemmed and hawed when Bob said that. It's got some implications."* Page's request caught Hale by surprise. Page had asked Hale to kick the foam strike uphill in the chain of command. Another of the NASA higher chain of command personages that added to the frustration of many at Kennedy Space Center was Linda Ham. Ham was the chairperson of the Mission Management team. Hale had passed the request of Page over to Ham. Ron Dittenmore didn't let them know that the request was from Page. Hale explained that the quality of the photos was not good and said the initial photo report with an estimate on the debris size, speed of impact and location had to be assessed. Hale said:[24] *"I did not pass on Bob's little comment about,"* Gee. Are there *some other ways to get photos? Remember the pictures are going to be on orbit. It's not what happened during the ascent. So, I didn't pass that*

---

[24] Comm Check –the Final Flight of Shuttle Columbia pg 72 M Cabbage/ Wm Harwood 2000, Simon Schuster

along to them at the time. Whether that was a major screw-up on my part, I guess that maybe it was." There was a four-day weekend that had to do with Martin Luther King Day. The NASA management team didn't meet until the next Tuesday, despite an ignored requirement on the books that they convene daily. Hale and most of the other shuttle manages would all but forget about Columbia's foam strike until then.

The shock settles in

Shortly after seeing images of the debris strike that morning, fill engineers at Kennedy Space Center, Johnson Space Center and marshal Space Flight Center began to spread the word of the potentially deadly hit on Columbia. A Boeing systems integration engineer at Johnson also looked at the film. He called a Boeing structure and debris analyst in Houston. The foam seemed to have struck the heat tiles on the Orbiter's belly near a corner on the leading edge. They never saw anything like this before. The JSC film team told the Mission Evaluation Room, (another team that NASA had in supply) that there had been a serious hit on Columbia. The Mission Evaluation Room, (MER) Johnson Space Center Flight controllers and MER engineers monitored the Orbiter's system at some forty plus computer consoles. An entry at MER daily log before 11:00 AM on Friday had this listed. *"{The Debris} travels down the left side and hits the left wing leading edge near the fuselage. The launch video review team at KSC thinks that the vehicle may have been damaged by the impact."* The entry also said: *"Two United Space Alliance employee deputy orbiter engineering manager Bill Reeves and TPS technical manager Mike Stoner were told."* United Space Alliance, a Boeing and Lockheed partnership held the shuttle's prime operations contract and would do the lead analysis. There were many opinions about the impact. Stoner sent out an email to other company managers just after 4:00 PM. Calvin Schomberg, a NASA tile expert and Mike Gordon, subsystem manager for RCC panels said, the tiles protect the ship from 3000 degrees of heat during re-entry were also notified.

Stoner's email:[25]

"Basically the RCC is extremely resilient to impact type damage. The piece of debris (most likely foam ice) looked like it most likely impacted the WLE (wing leading edge) RCC and broke apart. It didn't look like a big enough piece to pose any serious threat to the system. At +81 seconds the piece wouldn't have had enough energy to create large damage to RCC/WLE systems" The email noted that post analyses showed Columbia should be able to make it back safely even if the impact had penetrated the Orbiter TPS."

Stoner's message continued:

"As far as the tiles go in the wind leading edge areas they are thicker than required and can handle an area of shallow damage which is what this event most likely would have caused. They have impact data that says the structure would get slightly hotter but still be ok."

At Kennedy Space Center, the rumor the Shuttle Columbia would be home early was flying around the control room. It was standard procedure to bring an Orbiter home if there were major problems. However, an early return wasn't going to help an orbiter that was so fatally damaged. However, trying to make NASA go past its "Emerald City/Wizard of Oz" mentality was a job in itself. What was it going to take to make NASA see what those engineers in the film room saw? Columbia was doomed unless someone came up with a plan to save her. No one in NASA's upper echelon saw it.

---

[25] Comm check—The Final Flight of the Shuttle Columbia—pg 98 M.Cabbage/Wm Harwood 2000, Simon Schuster.

USA started to organize debris assessment teams that started some preliminary work. In the meantime, what about the request for photos from the Military? On and on, back and forth, more teams, more meetings, more distorted information, more time wasted. What was it going to take to make NASA see that they had a real problem that needed to be addressed?

There were so many people at KSC and JSC that knew the truth. Columbia was in trouble and yet no one wanted to use the one tool that they had which would help figure out what to do. Finally, Boeing engineers went to work and started asking some critical questions:

Where exactly did the foam strike the Orbiter?

- What was possible extent of damage to the Orbiter's TPS (tiles)

- How would the damage effect heating the orbiter as it re-enter earth's atmosphere

- How well would the ship hold up in the heat of reentry?

The piece of foam that broke off was 20"x16"x 6". It hit the Orbiter at the speed of 510 mph which raised the possibility of a more direct strike instead of a just a bounce off type of hit, which it seems most of NASA felt happened. The Boeing analysis team felt that the hit was along the black tiles between the left wing leading edge and the landing gear door. They were right on the money. That put the hit close to one of the twenty-two U-shaped reinforced carbon-carbon panels that lined the leading edge of the spacecraft.

Many engineers felt the RCC panels couldn't be damaged. There were only two incidents out of one hundred and thirteen mission, where a strike of that incidence happened. In 1991, the shuttle Atlantis was hit by foam as well as Discovery in 1999. Both had debris impact that hit the

RCC during flight. It was never explained and not considered of consequence by NASA.

A computer model program called "CRATER" estimated the hit would penetrate Columbia's heat tiles. That meant that the wing was vulnerable and the aluminum airframe could be exposed to blowtorch like hot gases during reentry. Using a 1984 formula, the program showed the RCC panels to be 0.246" when, in fact, it the panel's thickness was only 0.233" This was not good news for the Orbiter. This problem along with the uncertainty of the proximity of the hit, things were looking poorly.

The Debris Analysis Team continued to break down the information and found that the impact area included part of the main landing gear door on the left side. If the hit breached the door, super hot gases could invade during reentry and destroy the landing gear or worse. Unfortunately, it was much worse than expected.

At the Tuesday morning meeting, after the long holiday weekend, the Mission Management group met at the Action Center at JSC, led by Linda Ham. We should mention that Ham's Navy husband was in the Astronaut program. Mission control had offered her the job of flight controller when she first entered NASA's work force and many at NASA felt that she had moved too far up the ladder, too fast. As the NASA meeting progressed and the seriousness of the situation had really started to sink in, Linda Ham was sitting there wondering how this would effect the shuttle turnaround time and future missions. She didn't get the message that Columbia was in critical condition and might not make it home. There might not be a program to worry about.

The Shuttle managers and ISS managers were most definitely under the gun to get the job done of building the ISS. The job needed to be completed by February of 2004. The ISS had a deadline and falling behind would put major stress on the ISS program. Hence, NASA was between and rock and a hard place when it came to completing the ISS.

Don McCormack, the MER (Mission Evaluation Room) manager, had something to say at this meeting:[26] *"As everyone knows, we took a hit on the somewhere on the left wing leading edge and the photo/tv guys have completed, I think, pretty much their work, although I am sure they're still reviewing their stuff. They've given us an approximate size for the debris and an approximate area where it ... we're talking about what you can do in the event we have some damage there. "*

Linda Ham said: *"I'm not sure that the area is exactly the same where to foam came from ... but the [aerodynamic] properties and density of the foam wouldn't do any damage....I hope we had a good flight rationale then. "*

Obviously, Ham didn't get the big picture even though she was made technical assistant to Dittenmore in 2000.

McCormack says that he thought back to 1997 and a Columbia mission when there was damage on the tile between the wing's leading edge and the main landing door. He said he would gather the data from that flight.

"I really don't think there is much we can do. So it's not really a factor during the flight because there isn't much we can do about it."

Linda Ham responded: *"But what I'm really interested in is making sure our flight rationale to go was good, and maybe this is foam from a different area and I'm not sure and it may not be co-related but can you try to see what we have? "* After the meeting, Ham emailed Dittenmore about the foam rationale that the External tank manager, Jerry Smelser had some two months earlier on Endeavor's FRR prior to launch. Dittenmore said: *"You remember the briefing Jerry did it and had to go out*

---

[26] Spaceflight now—Mission Management team hardly discussed foam strike/ CBS News Space Place/ Wm Harwood July 22,2003—NASA transcripts

and say that the hazard report had not changed and that the risk had not changed... but it is worth having a look at again."

Linda Ham replied:[27] "Yes, I remember. It was not good. I told Jerry to address it at the ORR (Orbiter Roll Out review) next Tuesday (even though he won't have any more data and it really doesn't impact Orbiter rollout to the VAB. I just want him to be thinking hard about this now, not wait until IFA review to get formal action. From this Dittenmore email, McCormack emailed Ham all of the briefing charts that Smelser had and after looking at it, Smelser had nothing so different so based on that information, everything should be just fine. Linda Ham sent another email to Dittenmore: *"The ET rationale or flight of STS 112 (Atlantis) loss of foam was lousy. Rationale states we haven't changed anything. We haven't experienced any "safety of flight" damage in 112... So the ET is safe to fly with no added risk."*

The problem with all these emails is that they failed to get to the heart of the problem, that being what could NASA do to "see" the damage to the Orbiter now. It seems like everyone was so busy "checking six" a military term for watching your butt, that no one got to the real issue. Was Columbia safe for reentry or not? No one was able to answer that question. They all looked at other flights, which was looking at past mistakes.

The fact that foam debris was just as serious as O-ring issues was missed or to put it succinctly, passed over because... nothing had happened. The problem existed but because nothing had happened, nothing was done.

NASA continued to make the same mistake of looking the wrong way for an answer. The immediate problem of figuring out the Orbiter's viability to re-enter the atmosphere was the main problem. Yet due to the

[27]Columbia Accident Investigation Board –Chapter 6 NASA Decision making.

NASA bureaucracy, by the time they finally got around to figure out there was a problem…Columbia had burned up in the atmosphere.

More photo requests

The NASA engineers were concluding that something needed to be done regarding the Orbiter. The initial request for photos from the military was still no. However, on Wednesday morning there were three photo requests for Columbia in orbit that independently went up NASA's chain of command. Wayne Hale, NASA's launch integration manager, started to look at the emails that had been sent five days earlier regarding the photos. His assistant, Dave Phillips was the liaison to the 45th Space Wing. He asked the Air Force office about photo requests. The officer told him that a request had already been sent in. With two different requests for photos of shuttle damage, the USAF was there and ready to help in any way that they could. NASA did have a *"Memo of Understanding"* for photo requests with the National Imagery and Mapping Agency. The photo request worked for about an hour and a half. After talking to Phillips, Hale called Phil Engelauf, JSC's Mission operations directorate to ask for a formal request through channels. Austin called Linda Ham, to let her know. Lambert knew he should have gotten her permission first, but he didn't. Ham told Austin he didn't have the authority to request photos and who needed them? Austin told her, USA (United Space Alliance), Bob White and others. Ham started to call around. She called the Vehicle engineering office to she if they wanted photos, the answer was no. This judgment was reinforced after talking to the tile expert, Calvin Schomberg:

Ham said:[28] *"Calvin's discussion with me centered around the fact that the analysis that was being performed—he had seen the preliminary*

---

[28] Chapter 6 Decision making- Columbia Investigation Board / Appendix G

was—a worst case scenario, meaning that they assessed that the tile was gone completely down to the densified layer which is just a very thin tile remaining attached to the structure of the orbiter."

Ralph Roe, shuttle engineering director, told Ham: *"The conclusion that we came to in the discussion was well we are doing a worst case scenario analysis. We will let that analysis tell us what the worst case is and information from a photo. We didn't think would improve analysis."*

Linda Ham said: *"Wing leading edge, of course that didn't really even cross my mind. I don't think at the time. So, Ralph saw no reason for me to pursue the imagery. So I asked the MER and the MER said Well, no we don't know of anything. We will go check, too so I hung up the phone and they call me back and say well nobody we know is asking I said all right. Then I thought Loren Shriver of USA and I called him He searched around and he said he couldn't find out who was looking. So, I called Wayne back."* *"Apparently, Shriver found out during the calls that the request came from Bob Page and Bob White, but it was too late. Dittenmore and Ham already had agreed that the request for photos should be "turned off."* Linda Ham called Wayne hale to tell him:

Ham said: *"Now Wayne, we don't have anybody who wants a requirement. You know this is a busy flight....we've got all these science payloads and it carefully integrated and you've got to fly pointing this direction for the payload.... We don't know what we want to take pictures of, Every time we've taken images before, they have never been useful to us. Wayne, you know, even if there was damage there's nothing we can do about it. And it wasn't fatalistic, we're going to die, it was just that OK, so you came back and say we have damage, we don't have a tile repair kit, you know all this stuff about flying different reentry profiles which everyone has talked about ad nauseum, there isn't anything you can do about it. You just have to hope it holds together"*

So there we are, folks. NASA had Columbia dead and buried before she was even lost.

Linda Ham continued with: *"So you know, well let's just turn this off."*

Hale acquiesced with an *"OK."* Hale was very angry when he hung up the phone. He followed Ham's directive and cancelled the photos. Hale wasn't the only one devastated by the call. Meantime, the photo request had traveled all the way up the chain of command right to Cheyenne Mountain, Colorado and USAF territory. The USAF tried super hard to make these photos happen, but NASA just wasn't buying. Ham in the meantime, was trying to reassure herself that her call was the right one. Dittenmore told his assistant Linda Ham,[29] *"Another thought we need to make sure that the density of the external tank foam cannot damage the tile to where it has an impact on the Orbiter."*

That was good, critical thinking on Dittenmore's part. Ham decided to unload the problem on the engineers who still sent back a request for a photo, which of course, they never got.

No photos…no luck…no Orbiter

Let's reassess what we have here. A group of engineers was living in fear that the shuttle was in danger. The upper echelon of NASA, not wanting to take a hard look at the situation because it would require possible thought regarding some of their requirements, pooh-poohs the silly idea that the engineers had of trying to pinpoint the damage to the Orbiter by asking the USAF to take some photos for them. The USAF is willing and able, but NASA higher ups are not. Why question and see what the damage is if you can't do a damn thing about it? At least that is

---

[29] Academy of Program- Project and Engineering Leadership- Columbia's Last Mission- NASA- 2011

how it is presented in various phone calls and emails back and forth to NASA upper echelon. No one has really explained what is going on down on the ground to the astronauts who are busy doing experiments. An email was sent to give them a heads up in case any reporters asked in an upcoming interview about the foam strike. We wouldn't want the astronauts caught off guard on that one. And by the way, it's ok, not to worry. We've seen these phenomena before. We also have some engineers that are feeling an inordinate pressure to corroborate management's belief that all was well, and the foam strike was not a safety issue at all. One senior NASA official was worried that all this might cause a delay in the future flight schedule. Here we have a synopsis of the issue of Columbia. Simplified, yes, but it was right on the money.

There was so much confusion, back biting, requests unanswered, requirements shoved under the table and higher ups that should have lost their jobs on the spot for the crassness of how they handled their subordinates. None of that happened, however. It should have, but didn't. While some were content with their position on the subject, there was one man who brought something else to the table. John Kowal, a NASA thermal engineer sent out an email: *"In the case he ran, the large gouge is in the acreage of the door. If the gouge were to occur in a location where it passes over the thermal barrier on the perimeter of the door, the statement that there is "no breaching of the thermal and gas seals" would not be valid. I think this point should be clarified; otherwise, the note sent out this morning gives a false sense of security."*

A new thought was tossed into the ring. The thermal seal was critical to the protection of the Orbiter. The barrier is made of silicon with flame resistant Nomex coating. It would keep the super hot gases from getting into the well and damaging the landing gear or tires. Unfortunately, that is precisely what happened.

Frustration and Agony

At this stage of the mission, there were NASA engineers that were literally tearing their hearts out because they knew what they were thinking was true. The Orbiter was in serious danger and no one was really listening. It was as if there was a steel plate between the bottom line of NASA and the upper line of managers and officials. Many NASA insiders were furious at the insidious bureaucracy that just wouldn't quit. There were so many power factions at work here it could make your head spin and technically that's just what it was doing to the workers who knew something was drastically wrong and could do nothing to stop the worst from happening. In an email written by Kevin McCluney, a NASA flight controller: *"... There are only a limited number of choices (1) Do nothing assume it's just a bunch of smart transducer failures or that the gear can take the punishment. (2) Decide that the gear is probably toast, call for an early enough display to allow for a bail out if required (Less than Mach 0.9) and relies on the "remaining data or video in order to decide between a bail out and a landing attempt. (3) Decide that the gear is toast, that landing is impossible and call for a bail out. (4) Decide that the gear is toast and look for a gear up landing. #4 would mean losing the crew and vehicle, #3 would lose the Orbiter and only 1 and 2 are possibilities."* Columbia was into her second week of the mission. Above work went on normally, below chaos reigned supreme. Anyone ever hear of using the space station as a rescue vehicle? No, we didn't think so.

The photo issue just wouldn't go away. The USAF would later take the photos, all NASA had to do was ask. However, the war raged on, its possible not possible, chain of command regulations and requirements were holding the astronauts in NASA's hot, weak hands. As it looked, no one was going to stand up to find the truth about what happened and what to do about it. Columbia was prepared to come home the next day.

Columbia....on the way home?

There was an amateur astronomer and by trade, a defense contractor who just couldn't wait to get a glimpse of the shuttle on its way home. His name was Jay Lawson. He lived in Nevada and from his front yard; he set up to get his primo shot of the shuttle coming home. He made all his preparations and made sure equipment was set up to get a view of the shuttle. It would be a view that he would never forget, a view he never expected and a view that would haunt him for the rest of his life.

Lawson lined up the telescope and watched excitedly as he heard the sonic boom of the shuttle and saw a very bright flare. He thought that it was odd. He had seen something sort of break away from the shuttle. Lawson grabbed videotape that he was using to document the flight and plugged it into the TV set in his house. The picture he saw was strange and disturbing. What had happened? He turned the TV over to the NASA channel and couldn't believe what he was seeing. He started to videotape the channel.

Some 800 miles to the southeast, Kirkland Air Force Base had four satellites tracking specialists on duty. They were just outside of Albuquerque, New Mexico. The specialists had a little digital camera, which was attached to the telescope, which was looking at the computer controlled mirror system that was programmed to track the shuttle. The specialists were using the shuttle landing to try out some new software. It was a tough challenge to try to track a fast moving object with a large zoom lens telescope. The camera did exactly what they wanted; it took photos of the shuttle. You could see the Orbiter shape, but much of it was blurred due to heat generation of the reentry process.

Yet, the photos showed something happening around the left wing of the orbiter. There was some sort of vortices following the left wing.

Another witness to the shuttle's reentry was Jim Dietz, who was watching from his backyard. He had a 35mm camera and telephoto lens.

He watched, wondered, and snapped photos. Why was the shuttle so bright? What was all that material breaking away and trailing the shuttle? It hit him. Columbia was breaking up right before his eyes.

On Texas TV, the news was out and there was play by play as Columbia burned up in the skies overhead. Debris was falling like screams on deaf ears, the deaf ears of NASA and the lack of forethought or need to ask for some photos. This could have been prevented.

At Johnson Space Center while flight controllers waited to reestablish communications that were no longer there. As the Johnson control room started to lose sensor readings, the reality was dawning on them the Columbia was dying right before their eyes. The breach in the left wing had grown as super hot plasma gas flooded the Orbiter. The landing gear in the wheel well was burning up. The Flight director, Phil Engelhauf, looked over to Leroy Cain, the re-entry flight director. They knew as Cain shook his head, it was over. NASA had a new "contingency" and Engelhauf gave the word to "lock the doors". The word went out. *"This is mission control Houston"* James Hartsfield, Public affairs office at JSC, said over the NASA TV station. *"Flight controllers continue to seek tracking or communications with Columbia through Merritt Island tracking station."* The last communication with Columbia was at 8:00 AM. Central time. She was just above Texas and on her way to Kennedy Space Flight Center for landing. Flight director Cain was now telling controllers to get out their contingency procedures and start following it. By now, there were eyes all over Texas looking up and witnessing the demise of Columbia.

Back at KSC, the families and spectators were awaiting Columbia's arrival. At the end of the runway at Kennedy, Jerry Ross, veteran astronaut and director at JSC was told Columbia had not picked up on radar coming into Florida. He knew right away, the worst had happened. He left the Commander's Convoy Van, praying. Bob Cabana, Chief of flight crew operations at JSC, started to call out to the astronaut family escorts to the midfield-viewing site. Ross said[30], *"We think we've lost the vehicle. We need to get the families rounded up and sent them back to crew quarters as soon as you can. Don't say anything."* He called the crew quarter's staff and told them of the loss of the vehicle. *"Secure the facility. Get security out there. Turn off the TVs. Tell everyone what is happening. We're coming back and the families will be there as soon as we can."*

Laini Mc Cool was sitting up on top of the bleachers trying to hear what was being said on the loudspeakers. She and Jon Clark knew something was wrong. Laini McCool said[31]: *"I looked around and to my left I saw the non-family side of the bleachers, I saw men in suits, high level NASA as I was able to recognize a few, all putting cell phones to their ears. Soon everyone was on the phone. Then they started walking towards the parking lot, I ran down the bleachers to ask one of our family escorts what was going on when someone grabbed my elbow and told me we had to go. The whole event was surreal."*

Evelyn Husband and her kids, Laura and Matthew were waiting to hear the sonic booms announcing the arrival of Columbia. [32] *"We didn't have a clue what was going on. We were all standing then at the landing site, very light hearted. Matthew was running around playing. I asked what direction the boom would come from and out assistant had the most horrible look on his face. Laura asked if everything was alright and Matthew was absolutely silent."*

---

[30] Space.com Astronmaut Jerry Ross recalls Columbia Disaster, Feb 1, 2013 Clare Moskowitz, asst managing editor

[31] Comm Check—the Final Flight of the shuttle Columbia- Pg 154 M Cabbage/Wm Harwood/ 2000 Simon Schuster

[32] Comm check- The final Flight of the Shuttle Columbia –Pg154 M Cabbage/Wm Harwood -2000- Simon Schuster

Jon Clark, husband of astronaut Laurel Clark, said: *"It was like a president's been shot. You know all these guys are pushing you and getting you onto the bus or into those cars as fast as you can get out of thee. And nobody's saying anything....Just We've gotta go."*

The time ticked forward slowly on the big mission clock. The time was 9:16 AM and Columbia was overdue. At 9:29AM, Sean O'Keefe, the NASA Administrator had told Andrew Card, White House Chief of Staff, something was seriously wrong. President Bush got on the phone with O'Keefe at 9:45 AM and O'Keefe told him, "We have a disaster". The president asked where are the families and O'Keefe assured him they were being well looked after. It was a shame that O'Keefe couldn't have reassured the crew of Columbia that their spacecraft was in good order. Lack of images, chaotic management, frustration on the part of lower staff engineers who had the right information and the upper echelon staff that couldn't get it right if someone pounded the information into their ear, O'Keefe had a lot of explaining to do regarding NASA. Again, NASA had not listened to what was urgently needed, in favor of playing bureaucratic games. Another shuttle was lost along with her valiant crew.

Columbia had broken up over Texas. Seven more families experienced the most horrific pain of losing their loved one and were now in deep mourning on a day that should have been one of sheer joy. Debris was scattered over two states. Why? Again we will look at what NASA missed and why.

The Columbia Accident Investigation Board

The debris that rained down from Columbia was picked up in Texas and beyond. Again, NASA sat around a big conference table and talked about what, when and why. Who was responsible and how could this have happened?

It happened at 8:59AM. Columbia started her descent and re-entered over Texas. There was an increase in drag on the left wing. The flight system computers tried to compensate, much like they did with Challenger to correct the problem. It was in vain. Soon after that command, there was no signal from the Orbiter. Ron Dittenmore began press conferences every day to make sure all had daily updates. He told reporters that they might be able to extract the last 32 seconds of telemetry beyond where Rick Husband made the final call to Houston. They did have film of the shuttle breaking up just past 9AM.

It didn't take the press long to hit on the foam debris. They already knew about way before the astronauts did. Dittenmore answered the reporters with he was averse to jumping to conclusions because he might miss something else. He did say that the wheel well might have something to do with the disaster. He missed something. He missed the fact that the hit was brought to his attention and he fluffed it off. Dittenmore was questioned about not getting the close up photos from the military. He said NASA's previous experience with such imagery indicated it wouldn't be conclusive or provide in depth perception necessary to judge the severity of a strike. Dittenmore said:[33] *"The second factor was even if I had that information, I couldn't do anything about it. I'm helpless to go out and do tile repair. And the third factor was I had done the analysis. The best experts at our disposal concluded that it was a minor problem, not a significant problem. And when you added all that up, there was no need to take pictures to document any evidence because we believed it to be superficial and turnaround time issue… not a safety issue and so we didn't take any pictures."*

Based on Dittenmore's response, and based on all the prior information containing what other NASA engineers and managers thought

---

[33] CNN Rush Transcripts Feb 2, 2003 NASA Briefing

regarding the need for those photos regardless, Dittenmore felt the Orbiter was condemned and there was nothing he could do about it. If that isn't the most pathetic statement from a person who was paid to protect the Orbiter with his vast experience as an engineer, it was no wonder Columbia broke up. It was obvious Dittenmore was circling the wagons around his butt and shrugging his shoulders. Here we have the basic, rotten core of NASA, a core that needed to be expunged. The bureaucracy of trying to get those photos taken, the toes stepped on and the arrogance of the higher-level managers was disgusting at best.

What about Apollo 13? If it was up to Dittenmore and his crew of cronies, Apollo 13 would be on the way to the next galaxy by now and forgotten. The "Don't go outside the box" mentality of NASA is really so sad. We have all of these wonderfully talented men and women. People that were mid level in the hierarchy, even lower level or contractors. Those people cared more about helping the Orbiter and her crew than these highly paid, bureaucratic, highhanded engineers who, in the terminology of aviation, should have had their tickets pulled and showed the door.

Did NASA not learn anything from Challenger? Was that disaster forgotten? It was the same story with a different Orbiter. There were things that could have been done. At very least obtaining the photos would have bought some time to think about getting them home safe.

Had the damage been seen, blurry at best, it would have been one more piece of information to add into the puzzle. But because due to caring too much, some engineers went past Ms. Ham and her cronies, They were slapped down and reprimanded for even thinking about going out of the chain of command.

CAIB – Columbia Accident Investigation Board

We will now enter the realm of the CAIB. The panel came into being ten days after the disaster. It was headed by Admiral Harold Gehman. There was a long list of participants in the panel from Sally Ride to Neill Otte, a NASA engineer. Gehman had an impressive resume including the investigation of the bombing of the USS Cole. The partial list of panel members: Rear Admiral Stephen Turcotte, Commander, Naval Safety Center Maj. General John Barry, Director, Plans and Programs, Headquarters Air Force Materiel Command Maj. General Kenneth W. Hess, Commander, Air Force Safety Center Dr. James N. Hallock, Chief, Aviation Safety Division, U.S. Department of Transportation, Volpe Center Mr. Steven B. Wallace, Director of Accident Investigation, Federal Aviation Administration Brig. General Duane Deal, Commander, 21st Space Wing, United States Air Force Mr. Scott Hubbard, Director, NASA Ames Research Center Mr. Roger E. Tetrault, Retired Chairman, McDermott International Dr. Sheila E. Widnall, Professor of Aeronautics and Astronautics and Engineering Systems, MIT Dr. Douglas D. Osheroff, Professor of Physics and Applied Physics, Stanford University Dr. Sally Ride, Professor of Space Science, University of California, San Diego Dr. John Logsdon, Director of the Space Policy Institute, George Washington.

The first thing the CAIB did, was to take its offices away from and would not hold any press conferences on NASA property. The second thing it did was close ranks. Gehman wanted to make sure his witnesses and officers were not in the presence of anyone from Capitol Hill. Gehman wanted the truth and he wanted it now. He would not tolerate press leaks, lawmakers and their staff. The press and Capital Hill was given only a limited access to transcripts in a closed environment. He interviewed NASA personnel in private with their confidentially assured. The details regarding the photo requests, debris analysis and email traffic between engineers who were deeply concerned about the Orbiter were showing up in the press a week after the disaster. It was already known that NASA had major management communication problems and what was more deadly, NASA's handling of safety issues. In truth, whatever NASA had on paper, looked great. In reality, it stank like a rotten onion and that was precisely the way to describe it. The more you peeled away the layers, the bigger the stink got.

Gehman and the other board members were struck dumb during a March 25th Public Hearing at Cape Canaveral when testimony by Bill Higgins the KSC's chief of safety for the shuttle testified. In Higgins's testimony Gehman had asked Higgins three times exactly what it was that he did and three times, the man wouldn't elicit an answer. Gehman: "You said if [KSC Director} Roy Bridge wanted to give you some more people you would put them to work... I thought this was a shuttle program function and shuttle funded, in which case you should have said If Mr. Dittenmore wants to give me some more people. Or have I got that wrong?" Higgins: *"Well. You've got me on that one."*

Gehman continued to ask Higgins where the safety organization was in the NASA food chain. Gehman: "Where is that organization and where is that place in the food chain that we should be looking at?" Higgins: "That's a very good question. I'm not sure I can answer that specifically because that would be program and agency functions that are above me."

Gehman was not only frustrated but also truly shocked with Higgins testimony. Higgins said he didn't know whom he worked for. He contradicted himself and couldn't remember whom he was working for. Did Higgins not know or was it just a ruse. He either feared for his job or was covering something up. What this did boil down to was the Safety officer was not able to connect with the rest of the NASA centers if Higgins was to be believed. In truth, it wouldn't be a big surprise. Gehman talked about what the investigation had found.[34] *"These {staff} people started telling us stories of not very positive management techniques [if] somebody had a different opinion from the party line the they were intimidated or ridiculed or things like that...There is not one organizational trait that we complained about that we didn't witness ourselves. We were actually in the room, or we were there and saw a safety person do nothing, or we saw engineering being ignored and things like that. We actually saw it for ourselves. Slowly over a period of many, many weeks we came to the realization that there was more here than faulty foam. We started asking ourselves, how do we study and come to conclusions on just as much expertise in this social engineering area as we are in physical and thermal dynamics in the physical cause area? We said: We have to base it on a whole bunch of experts just like we have experts in the other areas. So we started searching for who those experts were."*

The board found their experts by the listful, but it could come to one conclusion: the similarities between Columbia and Challenger were blatant. If it had to do with risk acceptance, engineers who had issues that they would bring to the table and the issues were dismissed at the highest levels, we were right back at the same problem with Challenger. Bad communications, schedule demands and the fact that the press would not get off NASA's back when they needed them to.

The Challenger O-ring and the Columbia foam came to the same bottom line. Both these issues were going on from early n the shuttle program and worked their way into becoming an undisclosed risk. The trouble lay in the deep recesses of NASA's gut. The mid level managers had a hard time reaching the upper level managers like the request for the photos that just fell by the side because Ms. Ham and Mr. Dittenmore,

---

[34] Comm Check—The Final Flight of the Shuttle Columbia-Pg 202 M. Cabbage/Wm Harwood –2000- Simon Schuster.

refused to see the nature of the problem and decided they did not want to open another requirement and risk turnaround time for the program. Well, the program was down for two years due to the fact they didn't feel it necessary and seven people died.

Since Challenger and even before Challenger, the rule of thumb for NASA was **"Prove that it isn't safe."** Instead of prove it is safe. Many had forgotten the shuttle was not an operational spacecraft. It was still a research and development spacecraft no matter how much they wanted to call it a done deal. Just because the word was given to freeze structural, design didn't mean the shuttle was the perfect spacecraft. The space shuttle was never going to run like the "airlines" regardless of how much Public Relations wanted to sell that idea to the public.

Contractors have their say

Contractors also played a part in what went wrong with the shuttle. In 1996, NASA started putting contracts together under one major umbrella. It was now called United Space Alliance consisting of Lockheed and Boeing. This was done to save NASA some money, however that never happened. The contractors were dropping the ball. In a 1996 Columbia launch, a critical circuit clicked out during launch and halted the count-down. The investigation found: "the workforce has achieved a conflict-ing message due to the emphasis on achieving cost and staff reduction and pressures placed on increasing the scheduled shuttle flights as a result of completing the space station. The investigation also found dozens of near misses that could have caused failures that only the hand of God prevented. However, the investigation did too little too late. Its effect didn't last long.

Because NASA was losing jobs due to attrition, retirement, what have you, its expertise level was also losing speed. Many of the contractors didn't want to leave their homes to transfer to another place and basically, the younger people moved in. The younger staff came in, but their expertise level was low. Hence, the workforce expertise level dropped pitifully. To this statement, Ron Dittenmore, Integration manager for all shuttle operations said; *"That skill base migrated over to the contractors. That included not only the bottom level engineering, but also the mid level management and organization lead management of these engineering organizations. NASA had to rely on the contractors input, involvement and their management position. Certainly, we would challenge their decisions, but we had to rely on their involvement. I find it interesting that as we go through this part of the FRR and the Mission Management Team that there is very little contractor management conversation or involvement I could see.... I don't have any management on the contractor side coming forward and telling me, things aren't right. I have only silence...when a chunk of foam comes off, it's the contractor's job to analyze that and tell whether it represents a problem. The contractors have the skill base. When the contractor comes forward and says we've done our job and we don't believe there is an issue, they become part of the system and they are accountable for that recommendation... and so when the board is critical of conversations that did or did not come forward to program management on their [government] side, I wonder where the contractor management was..."* According to Dittenmore, he really didn't need to have any part in any of these proceedings, the contractors should be held accountable, not the higher management of NASA who ran the show. You don't need a sociologist to analyze that statement. It's straightforward. Dittenmore wasn't responsible for anything other than picking up his check. The president of USA (United Space Alliance) wasn't taking those words too kindly. Mike McCully was a former shuttle pilot and now the president of USA said: "I disagree completely. We were not silent. We brought issues to the table. Our management teams support our technicians and inspectors when they found things that resulted in the grounding of the fleet.... We make heroes of folks who find things that ground the fleet. It's our management that does that."

So here, we have NASA and its contractors at each other's throats in the middle of the CAIB investigation. Gehman and O'Keefe were also doing a little throat biting over the fact Linda Ham and Ralph Roe, both pivotal position managers, who were working within the CAIB and investigating themselves. Gehman wanted then to step away from the CAIB and he wanted it now. After a bit of a struggle, O'Keefe accepted their "request" to remove themselves from the CAIB. It wasn't too shortly afterwards, around April 19, 2003, that Ron Dittenmore would also take the proverbial powder and leave the shuttle program. Later that summer he moved to an industry job. By July 2, Linda Ham's position was eliminated. She was changed to an assistant in engineering position to the engineering director, Frank Benz. All of this moving around was deemed to make it look like NASA was straightening up the house. However, there was much more to it than just that. Ham was being trashed severely by the press as being dictatorial bureaucrat who only cared for her career.

The CAIB digs deeper

As the CAIB went deeper and deeper under NASA's working guts, it was obvious that communications between everyone, contractors, internal offices, engineers and managers was so bottlenecked with bureaucratic nonsense that no one had a clue what the other side was doing. One group was ready to throw the other under the bus, no questions asked. The CAIB was fed up with O'Keefe and NASA's response to what the committee requested of them. Initially some NASA Shuttle engineers just didn't bother responding. Does the sentence "Open Rebellion" say something to you?

The CAIB was asking some very tough questions from NASA people who felt that they were being interrogated. However, there were some things that were never mentioned like Bill Readdy the Associate administrator for Spaceflight in Washington Headquarters, handing in his resignation to O'Keefe right after Columbia's failure to come home. However, did moving people around, firing them, accepting resignations do any good at all? The suffering they all had to face knowing that they were part of the disaster will never leave them. What could erase the memory?

As we continue to look for answers, we have to look at the International Space Station. The pressure was on to get the ISS completed and that schedule pressure found its way down the line right to the heart of the shuttle missions themselves. Hence, this too added to the lack of communication and internal strife within all of NASA.

The Rogers Commission had requested during the Challenger investigation that management be moved back to Washington, D.C. headquarters. In this case, the CAIB was also asking for the shuttle management and the ISS to be transferred back to Washington headquarters. However, this brought on what was secretly well known by all, the turf wars between all the centers that had been going on since the inception of NASA.

Sean O'Keefe has tried to put a bandaid over things and brought in a new manager for JSC in Houston. Air Force General retired Mike Kostelnik. He became Dittenmore's manager. Kostelnik was all Air Force and he was a tough nut to crack. He had made it known, loud and clear, he was the new boss and it was obvious Dittenmore wasn't having it, which many feel caused his resignation. What O'Keefe wasn't getting was this: Every NASA center had its own personality and culture. It has been like this since day one. O'Keefe's try at sorting out JSC with a new top dog wasn't going to work. At least not with the CAIB. Trying to take all the field centers and turn them into one cohesive NASA was so much science fiction. It just wasn't going to happen.

Let's look at another issue, the former NASA administrators that brought us to this disaster. Remember, Columbia didn't happen on its own. This was building up for years. Before Sean O'Keefe, there was Dan Goldin and he had a long track record. He was brought in by Bush 41, went through the Clinton Administration and that meant from April 1992 to November of 2001. He had the longest term of any NASA administrator. Goldin was responsible for many of NASA's current problem. The financial books were badly screwed up during his administration. The ISS was totally over budget. Goldin scarified he shuttle safety by shifting funds from the shuttle safety program over to the ISS. The faster, better cheaper, approach Goldin tried to sell to Congress virtually tore NASA's guts up. All it gave NASA was more robotics, which meant more money speak indiscriminately and it didn't leave any room for success. It just ended up in technical issues that were not resolved: two Mars missions that failed. Goldin was known for his rough temper and stunning scream out sessions with personnel. He earned the title given to him of "Captain Crazy"[35]. He was enamored by all of the fake awards he made sure he received and made sure his image was plastered all over the NASA Centers. Goldin was too long in the office but by the time he left, he reduced NASA to inner turmoil the only reason we didn't see Goldin in Congress fighting for funds was that he feared for his office. The CAIB wanted to interview Goldin but he refused to talk to the committee and would not allow them to record anything said. Thus, we have Goldin's support in allowing the CAIB to find out more if the deeper things going on in NASA, caused by Goldin's administration.

---

[35] P.221-Comm Check...Michael Cabbage/Wm Harwood; Simon &Schuster/2009

Inside the CAIB report there is an interesting selection called "5.4 Turbulence in NASA hits the Space Shuttle". Goldin engineered not one but two policy changes. This was not evolutionary change but radical or discontinuous change. The report goes on to state, "His tenure at NASA was one of continuous turmoil of which the space shuttle program was not immune."

A point brought out was that Goldin felt "Corporate headquarters should not attempt to exert bureaucratic control over a complex organization, but rather set strategic directions and provide operating units with the authority and resources needed to pursue those directions. Another deeming principle was that checks and balances were unnecessary and sometimes counterproductive and those carrying out the work accept primary responsibility for its quality. It is arguable whether these business principles can readily be applied to a government agency operating under civil service rules and a politicized environment." Goldin felt this principle was good for NASA. It wasn't. It might work for a tool and dye company but not for an organization as complex and seated in bureaucracy as NASA.

To add to the drama, Goldin's plan was that headquarters should worry about strategic issues. In 1996, Johnson Space Center, Houston was made the "lead center" for the shuttle program. This position went back to prior Challenger disaster. This was part of a move for all program management responsibilities from NASA headquarters to NASA's field centers. To add oil to the fire, Johnson Space Center would now have authority over the funding and management of shuttle activities at Marshall Spaceflight Center, Alabama and Kennedy Spaceflight Center. Johnson Spaceflight Center and Marshall had been internal rivals since the Apollo Program, and Marshall employees who were around for a long time, weren't taking JSC's control too easily.

Problems were springing up and it was obvious war would break out soon. The head of the shuttle program, Bryan O'Connor felt the transfer of management function at JSC would put things back to the same position before Challenger's demise. O'Connor said: "It's a safety issue. We ran it that way (with program management at headquarters) as per the Rogers Commission for ten years without mishap and he didn't see any reason to return to pre-challenger days." Regardless of the agreement, Goldin changed the protocol.

Goldin was also in love with the idea of privatizing the shuttle program. In 1994, NASA thought about consolidating many of the number of shuttle operations contracts to a prime contractor. There were eighty-six separate contracts held by fifty-six contractors. Here is where NASA through they would save some money. Famous, early space flight veteran Christopher Kraft recommended in his report of March 1995, that the shuttle was a *"mature and reliable system"*. This was wrong at the first take; the shuttle was a research and development vehicle and never truly became "operational". Kraft went on to say: "If NASA wants to make more substantive gains in terms of efficiency, cost savings, better service to its customers, we think it's imperative that they act on recommendations…We believe that these savings are real, achievable and can be accomplished with no impact to the safe and successful operation of the shuttle."

With all due respect to Chris Kraft and his service, never could have he been more in error. Moreover, the Aerospace Safety Advisory Panel said just that. The idea that NASA could further reduce the number of civil servants working on the shuttle brought KSC engineer Jose Garcia to write a letter to President Bill Clinton.

The letter from Garcia brought some very serious points to the table. The misguided attempt to economize lost direction when it came to keeping the shuttle safe. Garcia also said before Dan Goldin's tenure as

NASA administrator, Garcia felt that he could bring any point up and management would solve the problem. At the time of the writing, Garcia felt he now couldn't voice any concerns and that there were many engineers and managers who were leaving to program because of the fear of retribution. Garcia felt that it was impossible to have any kind of exchange of ideas within NASA, as he knew it then. At Kennedy Space Center, Goldin told middle management: "I have all my people in all the right places and anyone who doesn't live up is gone." Garcia also brought up the fact that the checks and balances system of processing the shuttles was in jeopardy. NASA had two teams processing the shuttles, the contractors and the KSC folks. This system crosschecked each other to prevent any catastrophic error. While this system was expensive, it was effective. Since everyone in NASA felt that anyone who doesn't have a higher agenda or in fear of losing their job would admit that you can't delete the checks and balances system of shuttle processing without effecting the safety of the shuttle and her crew. Garcia went on to point out that there should be no change in hands on processing of the shuttle. There are no second chances of getting it right. The very last place that cuts to the shuttle should be made is in the hands-on processing done by NASA and contractors. There have been first place cuts made during that past year and it is now the object of *"an unconscionable attempt to drastically change the checks and balances system which has been the backbone of safe manned space flight."* Garcia went on to state that Goldin's background is in unmanned space vehicles, and he did not understand all the requirements of shuttle processing. Another issue was privatizing the shuttle, which was a national resource. The shuttle belonged to the people of the United States, not to the contractors. The shuttle is a complex research and development vehicle, which you can't privatize and run it like a cafeteria.

Therefore, to take NASA out of the equation and give it to the contractors would be detrimental to the shuttle and the program. Garcia was adamant that KSC was the heart of the shuttle processing and maintained the responsibility for her safety. At this time, NASA was in the process of declaring the shuttle operational and giving pre-launch activities over to the contractors just before the Challenger disaster. It was a mistake then and is a mistake now. The shuttle is a research and development vehicle and should be treated as such. The environment that the shuttle was being exposed to could be causing some serious effects that even NASA wasn't sure of at the time. Garcia also brought out that 50% of the turn around processing work at KSC is unplanned. Removing NASA from the day-to-day activities and making NASA responsible for only auditing functions would result in NASA processing the shuttle in the same manner that unmanned launch vehicles are processed. Shuttle processing has already been downsized some 30% without significantly jeopardizing shuttle safety. Garcia pointed out that for fear of job loss, many were climbing on Goldin's "Bus to Abilene". KSC/NASA contractors launch teams are in risk management. Drastically changing the KSC launch teams will increase risk to the shuttle and her crew. If drastic changes were implemented, the perpetrators should not only be held accountable for their action, they should be criminally liable for consequences. Garcia closed out the letter with the KSC shuttle team was the best in the world and Goldin was trying to dismantle it. He was again adamant that NASA was running out of time and they need help to stop the madness. This threat by Goldin was the biggest shuttle safety concern since Challenger. It was a tremendous letter that Garcia wrote to the president, but it didn't do much for keeping Goldin's fingers out of the safety budget for the shuttle. One of the reasons, the foam issue was not pursued as it should have been, along with other shuttle safety upgrades was, Goldin used the money to fund the Russians part of the ISS, since the Russians were broke and couldn't pay for their partnership in the ISS. Hence, it was obvious why Goldin did not wish to speak or appear on record for the CAIB. He had a lot to answer for and he wasn't about to admit to any of it.

Let's take the privatization issue and look at that for a moment. In August of 2001, NASA headquarters had the White House "Privatization White Paper", that called for transferring all shuttle hardware, pilot and commander, astronauts and launch operations to a private operator. The CAIB did look at this issue. In September of 2001, Ron Dittenmore, shuttle program manager, put out his report the "Concept of Privatization of the Space Shuttle" This paper argued that, for the shuttle to "remain safe and viable," it is necessary to merge the required NASA and USA (United Space Alliance) and it was not optimal and it was unlikely that NASA would ever recapture the shuttle responsibilities that were transferred in the Spaceflight Operations contract. Dittenmore's plan wanted to transfer 700 to 900 NASA employees to a private organization, which included astronauts, flight crew, program project management, shuttle main engine, External tank, the redesigned SRB, mission operations, including flight directors and controllers, ground operations processing, launch directors, process engineering, flow management responsibility for safety and mission assurance. After all this happened, according to Dittenmore, "the primary role for NASA in shuttle management will be to provide SMA (Safety and Mission Assurance) independent assessment, audit and surveillance technology. All of this was under scrutiny at the time of the Columbia demise. It sounds like all Dittenmore wanted to be responsible for was his paycheck.

It took NASA until the end of the decade to realize Administrator Goldin's force cut had gone excessively far. By 2000, NASA needed to rebuild the infrastructure of its workforce. The downsizing led to services issues concerning skills on an already overtaxed workforce. As more employees left, the workload on the remaining force increased reducing operational safety and capacity. NASA needed to start hiring several levels of workers. Thanks to Goldin's "faster, better, cheaper" policy he put the wrench in the works of safety for the shuttle program.

What the CAIB said

There were a number of misconceptions that were brought on by the press and NASA during the CAIB process. There was a catch phrase incorporated into just about everything hitting print-- "Broken safety Culture." It was overused and what did it really say? It said NASA had serious internal issues regarding safety and communication. While the CAIB criticized NASA for its "safety culture", it went further into NASA's internal organization in total. NASA was unable to face the technical issues that caused Columbia and Challenger. There was also the issue that NASA never truly looked beyond the shuttle for the next generation of vehicle while managing an already problem beset manned spaceflight program. The CAIB made some serious statements regarding the arrogance and the problem of allowing managers to speak their minds regarding safety issues without the thought of retribution or loss of job.

The CAIB report looked back to the Apollo program that brought about the theory enforced of the "perfect place, alone in its ability to execute a program of human spaceflight." The CAIB report went on to state:[36] *"The NASA human spaceflight culture manifested in particular a self-confidence about NASA possessing a unique knowledge about how to safely launch people into space."*

The CAIB also found NASA to be a place of "cumbersome organizational structure, chronic understaffing and poor management principles." Management routinely "deferred to layered and cumbersome regulations

---

[36]Space Watch: A cultural change at NASA—Robert Zimmerman, March 2005

rather than the fundamentals of safety." It went on to express that "NASA's Apollo era research and development culture and its prized deference to the technical expertise of its working engineers, was over ridden in the space shuttle era by bureaucratic accountability—an allegiance to hierarchy procedure and following the chain of command." What all that says is that NASA played by its own rule book, whether the rules were up to date or not, more often not.

The CAIB really rocked the house with this statement, "that NASA's management often operated outside its own rules even as it held its engineers to stifling protocol." That was obviously very true when you consider what happened with trying to obtain photos of the shuttle in orbit. Those who went out of the chain of command were chastised, yet the higher level of command swept the problem under the rug for want of bringing up a requirement that would need explaining.

Flawed decision making on NASA's part came forth from its own self deception and that is something that NASA did do well, self deceiving itself, and lack of curiosity as to why things were happening. NASA played by the food chain rule and never really tolerated anyone stepping out of that realm. NASA allowed its management to insulate itself and look back on its halcyon days of "CAN DO" attitude, which made NASA higher ups unable to deal with the reality of problems. The CAIB report made it clear that without careful and thoughtful restructuring of NASA's management, it would find itself hard-pressed if not impossible to move successfully into the future.

What would NASA do now?

Even after NASA returned to flight from the Columbia disaster, the arrogance and unrealistic management practices never really changed that much. We even look, right now, at President George W. Bush's 2015 initiative to "return to the moon." To help formulate the plan, NASA advocated ending the shuttle program in 2010. This rather tossed out the baby with the bath water, as NASA had no form of back up to space flight. The first manned flight in the "Crew Exploration Vehicle" was slated for 2014. This left a four-year hole and no way of getting and retrieving astronauts from the ISS. Things got worse for NASA when the Obama administration came in and canceled Project Constellation in total, which completely diminished the manned spaceflight program for NASA and the U.S.

What was NASA going to do to keep its assets protected at the ISS? They made a deal with the Russians for $70 million a pop to pick up and send our U.S. astronauts to the ISS— This agreement was only going to be good up till 2019. Even worse than that, NASA made the decision knowing it would be forbidden from paying Russia for the use of the Soyuz capsule under the terms of the IRAN non- proliferation act. NASA's plans of replacing the shuttle with the CEV were not the best. In order to manage the contractors building the CEV, although it had been canceled, they had to submit one hundred and twenty-nine quarterly reports and monthly reports that was nothing more than an exercise in wasting time and money.

Nothing had changed since the CAIB. NASA still relies on its "Bureaucratic accountability" which the CAIB took them to task for. Instead of allowing the contractor to work and design, NASA decided to pile on the paperwork with the concept that NASA would review the materiel, written in the most boring of details, to see if the work was done correctly. To top that off, none of these reports had anything to do with the safety or good workmanship in engineering. The bureaucrats made sure there was enough paper to cover their butts. Reports such as these do nothing more but clog up the flow of information, waste precious time and energy in trying to divulge what these redundant reports are all about. The reports distract everyone from what really was important.

The reports cost money, slow development and hamper thinking, yet NASA is so reluctant to let these tomes go, they would risk more instead of just listening. Some things did straighten up for NASA after the CAIB investigation. Some of the NASA offices were sorted out and direct lines of action made more accessible, at least for now. NASA required a massive amount of "spring cleaning" after the Columbia disaster and while some of that may have occurred, not enough of it did.

The CAIB could not address other factors, like Congressional interference and port barrel projects that support the Congressional states and constituents. Money that NASA could have used for important projects was being used to support some Podunk project because senator so and so need the work for his state. There is much more that could have been accomplished with the CAIB report and it wasn't the fault of the panel. NASA had to be willing to change. While some things may have been corrected, the main meat and internal workings of NASA are still as convoluted as they were before Columbia.

New administrators have been in place since Sean O'Keefe left. Mike Griffin, who was an engineer, understood what NASA needed and he wasn't afraid to speak his mind. He did, on many things. However, Griffin didn't last long enough to make the changes needed. As to Charles Bolden, the most recent administrator, he is a total political appointee, sent in by the Obama administration to further dismantle the manned space program and NASA. Fortunately, Jim Bridenstine a former Navy pilot and Congressional Representative for Oklahoma has since replaced Bolden as NASA director.

Where we are after the CAIB

At the end of the day with the CAIB we find ourselves pretty much in the same place as after Challenger, lots to fix but not enough commitment to do it. The CAIB found many of the same problems that were in place with Challenger. It hadn't changed and now with the shuttles out and in museums, just what is left for NASA to do?

What did we really learn from the CAIB

There were a number of things that NASA needed to come face to face with. Much of it had to do with leadership and how it should handle problems, small, large, miniscule and humongeous. When a team leader has a problem brought to them by a worker, whatever the worker's level, it needs to be addressed, It needs to be looked into and satisfied. An answer needs to go back to the worker and the worker told, yes, this is a major issue you are right, or it's been taken care, whatever the answer, it must be the truthful answer. No one, at any level of work should ever be shunted aside because they aren't in the leadership chain of command. NASA needs to learn that everyone is important and everyone, what ever their level of expertise, should have pride in what they are doing and be respected by leadership for that pride in doing their job well. Problems shouldn' filed in a draw or pushed up the paper chain. They need to be addressed. NASA didn't see fit to do that. Because of fact that leadership refused to address the issues that workers did bring up, and also failed to address problems that they knew were dangerous or could cause damage to the Orbiter, 14 astronauts were lost along with two shuttles. NASA had two major problems; the first was their scheduling, and it was a mess of pressure and publicity due to certain "events" that were happening in both flights, the facts that showed they should NOT fly were not addressed. In the case of Challenger, the cold and ice, the fact that there was blowby shown before in the O rings, the fact that the contractor tried to show the problem with the O ring and was dismissed in the meeting because he didn't have the right "presentation" boards to show NASA and others involved in the launch decision. The second fact of the "teacher in space" which added extra pressure to the launch decision, was also in play as well as the lackadaisical attitude about missing a perfect day to launch due to a holiday. NASA attempted to correct some of those problems after the Rogers Commission, but not all. The major problems of bureaucracy, lack of leadership in addressing problems or addressing them too late came back to bite NASA once again with Columbia. A problem with the foam breaking off the External Tank was accepted as a "flyable" situation. It was a situation that happened in some 73 of 113 missions. The only missions that could not be counted were the night launches because the cameras could not see what was happening. Apathy of a sort settled in to just "accept" the problem, till January 16, 2003 when the foam broke off and the shuttle Columbia flew right into it, hitting the left wing and punching a hole in the heat shield. She was doomed as she re-entered the atmosphere to return home on Feb 1, 2003. This problem was a known factor, just as the O ring was. There is a very descriptive term for what NASA was doing in both these cases and it is called "Normalization of Deviance".

In essence, this term describes what NASA was doing. NASA was allowing both these problems to become part of the daily routine when it came to preparing the shuttles for launch and mission. We all found out it wasn't the right response for either issue.

NASA has been working on a threadbare budget since the shuttles have been retired. They are being asked to once again come up with the next piece of hardware to get the United States back into space, manned spaceflight to be accurate. Will the same things happen with the CEV? Missions that are going to the Moon and Mars will not tolerate "normalization of deviance". That will not work when you are sitting on the lunar surface on a moon base or millions of space miles away on the ride to Mars. NASA needs to have its ducks in a row and come up with a way to make sure apathy, problems needing to be addressed, and leadership are those ducks. There is no room for any type of normalization of deviance, not in a lunar mission or a Mars mission or in the construction of the CEV. NASA needs to find its way back to doing things the way that gave them the ultimate success in missions like the Apollo program.

Looking at how Apollo was done may be a good way to start. As nothing is perfect, NASA needs to make its missions and it equipment as problem free as possible. Yet, if Congress doesn't come up with the money and the internal workings of NASA when dealing with large programs doesn't change, what have we to look forward to?

The Manned Spaceflight Program

Where does the Manned Spaceflight Program go from here?

After exploring the beginning of NASA, watching it evolve into what it has become today, featureless, albeit, there are some magnificent robotic programs, but without the bold face it once had on the world stage. NASA sits on a lonely launch pad, with no way to the stars, let alone the space station. What plans does NASA really have left? Possibly, with the new Trump Administration Directive there might be a chance. Should NASA attempt to use its meager resources to chase an asteroid ala the Obama administration? Should it stay only with robotics that they have successfully landed on Mars, and explore via a little robot? Alternatively, should NASA dare to hope to reach out to the frontier of space? As it stands right now, NASA's meager 19.1 billion budget for FY 2018 isn't going to buy it much of anything. It is actually 3% less than the 2017 budget. There is not even a hope of redeeming its amazing manned spaceflight program, at least not with this amount of money. NASA remains embroiled in its bureaucratic nightmare, closing door after door on its remarkable past and manned spaceflight for the United States. Other nations have already passed NASA on trying to return to the moon, albeit, it might be that Nation's first attempt, that meaning China. China now has a temporary space station that they are maintaining quite well, thank you and most likely built on U.S./NASA concepts and plans.

After examining all of NASA's history including the Shuttle program that with the right leadership, Congressional care and presidential want, could have been even more glorious that it was. However, because of the fact NASA is at the whims of whatever and whoever holds the purse strings. Many of the things that should have happened, didn't.

Back in the APOLLO days, when NASA could do no wrong and would accumulate the funding needed to match the challenge, NASA proved what it could do and did do. In the 1970s with civil unrest and a Vietnam war that just wouldn't go away, President Nixon signed on for the partial program that included the shuttle without having a place for the shuttle to go, meaning the space station which would come much later. It was a beginning that was never really serviced that way it should have been. All was given to the lowest bidder. In some cases, that was fine, in other cases, it cost the program more than just money. The dreams of a two stage, fully reusable/recoverable shuttle with some 100-150 flights a year went the way of the wind. The hopes and plans for a nuclear stage to reach for Mars were canceled in 1983. Budget constraints choked the program as badly as the "Tower of Babel" bureaucracy that wormed its way through NASA like a parasite. As the shuttle developed via various stages like lifting bodies and the X-15, the military made sure that it took its piece of pie when it came to staying the leader in what the USAF felt was their sole domain, the sky and space beyond. The shuttle developed but along military lines, not commercial or civilian lines. The evolution of the shuttle requirements increased the payload bay for the military satellites that would need to fit. The cross range which was always an issue was determined by the USAF. By 1972, Nixon decided to finally go with the shuttle. The bottom line of what the Nixon administration proposed was:

The requirements:
- Five orbiters
- Reusable Orbiter and engines
- Reusable solid rocket motors
- Expendable external tank
- 40-50 flights per year

- 10-15 people per flight
- 5.2 billion and 20% in reserve for research and development.

Here's the catch, the shuttle never got out of the research and development stage. As much as NASA and everyone else tried, NASA never got out of that stage and that was one of the causes of the disasters. Everyone was asking for something that just wasn't there. The shuttle was not and would not be transformed into a fully readied project. It was a research and development spacecraft and would remain so till the last day of flight. Why was the shuttle considered an R & D project? Basically because her purpose had never really been developed. As stated before, when President Nixon laid out his plans for the shuttle, there was no mention of the International Space Station, that in fact, was an after thought. Not so much an after thought, as a NASA could not have its cake and eat it too. It would have to be one or the other, Space shuttle or space station, NASA chose the space shuttle. At that point in time, the development of the shuttle was something that had never been done before, it took massive amounts of work to create this magnificent and complex spacecraft. That work was still on going right up till the last launch.

Was the shuttle program a failure...no. The program was a pure and unadulterated success. Yes, there was loss of life. The reasons for the loss of life could have been stopped. However, due to the issues that were in play at the time, it was inevitable that it would happen. Sadly, that is so true, and it was endemic to the process NASA was involved in. NASA couldn't see the forest from the tree. Many goals weren't reached. One of them had to do with 50 flights per year, NASA never even got close to that. At something like $200 billion a flight, it was unreachable. Also many felt the shuttle was limited in its reach. That is true. The shuttle could attain a low earth orbit, and that was that. There was never any hope of going farther than that. However, we have to credit the shuttle with saving the Hubble telescope. What a mission that was! It is the shuttle's proudest achievement. The construction of the International Space Station was another massive achievement for the shuttle. However, because of mismanagement of funds by a NASA administrator, the shuttle's safety was compromised. Right now, the Russians control the U.S. space program and the ISS, as we must pay them $70 million to ferry astronauts back and forth. That expensive ride will end in 2019 and while astronauts have been chosen for the new Orion Multipurpose Crew Vehicle. It may be years before anything transpires to a launch. Can we really call the ISS or even part of the ISS our own?

The ISS

That is another point, the ISS. While the ISS is a success in many respects, what have we accomplished there? Yes, many experiments have been run, however, many of these experiments have been run over and over "ad nauseum". What is to be truly learned from them. Have they advanced our knowledge of how the human body reacts to prolonged time in space or is it just a "tourist" attraction the Russians use to get money, by selling a ticket to the station as a "payload specialist" to some rich millionaire with nothing else to do? Yes we do have some record holders for the longest stay at the ISS. Scott Kelly, stayed on board for 438 days, while two Russian astronauts stayed for a combined 878 days.

We did learn something about the effects of long term living in a zero G environment. However, what are we going to apply all this science to if we have nowhere to go. The Trump Administration has signed the directive to put us back on track to a lunar base as a jump off point to Mars. But, will it really happen?

What happens now?

What does NASA have to do to take itself back to its former glory and that does not mean egotistically, it means the engineering/mission management that it once handled so well. NASA needs to refocus itself on certain goals. One of the first things it must concentrate on is to decide on its place in space exploration. How can it achieve the main goals instead of settling for the smaller, short term missions that gain very little. NASA also needs to determine how and where it gets its funding. How money is allocated to NASA needs to be revamped via Congress. NASA needs to walk into Congress with a viable plan, not a pie in the sky dream. NASA has lost a tremendous amount of its finest talent. Astronauts that have spent years training, have walked off the NASA payroll because there was no hope of flying and no work to be done except for publicity. There is only so much training you can do if you have nowhere to go. NASA also needs to rebuild its internal work force. Instead of the cheaper compartmentalized worker, workers need to be trained in many parts of a project, not just the one nut or bolt they will screw in. NASA needs to bring America back into the space program. It needs to fly the NASA meatball flag next to the USA flag and say, "Let's go and do it!" NASA needs to have itself out there publicly, in the grammar schools, high schools and colleges. They need to give the youth of the USA something to shoot for, something to make the youth of the USA proud and willing to work those math and science courses that will land those NASA jobs. It doesn't mean all math and science, NASA must emphasize man's reason for going into the final frontier.

NASA needs to stop the bureaucracy and look at what is directly ahead of it. NASA needs to leave the many tiered, useless layers of bureaucracy behinds, streamline, make it easy for a worker who sees something to get to a manager and talk to them without fear of stepping out of the chain of command. That is part of NASA's current problem.

The Bush Administration had the right idea. The Bush Administration looked again to the moon for a landing in 2020 to reestablish a base there as a jumping off point to other planets like Mars. However, the moment the Obama Administration got into office, it canceled CONSTELLATION that would have been the new crew vehicle and heavy lift rocket. While at this writing we are still fighting over a piece of paper that represents our heavy lift vehicle that will never be built as long as the Obama administration was in office, NASA needs to have the leadership in the front office to push the programs through and fight for what NASA needs.

Politics and cronyism has destroyed NASA from the inside out. Presidential appointees like Charles Bolden, who is one of the weakest NASA Administrators in history and a political crony of Obama, has shown just how this can happen. At one point Bolden was an astronaut that flew on the shuttle. When the shuttles went into retirement and were sent to various museums, it didn't take much to understand Bolden didn't have a clue as to the significance of the Orbiters to aerospace history. When you look at the decision making process used by Bolden, it was laughable and frightening to see how he handled these most precious artifacts.

It was a travesty and one Orbiter, Enterprise, was even severely damaged due to being on the 17th story of an aircraft carrier museum, in a bubble tent. It wasn't long before Hurricane Sandy took care of the bubble tent and damaged the Orbiter. No thought, at least sentient thought, was given to the dispersement of the Shuttles to various museums.[37]

Nor did Bolden give much thought to NASA programs, other than rolling out a statement in an Al Jazeera interview for NASA to reach out

---

[37] NASA and the Shuttle Shuffle- The Disposition Process for the Orbiters and how it all went wrong - 2012 Phoenix Aviation Research—Jeannette Remak 2012

to the Muslim world.[38] This wasn't what NASA needed or wanted to hear in light of retirement of the thirty-year shuttle program which was haunting all of NASA.

NASA revived?

NASA needs to seriously revamp its goals for the entire organization. The Space Foundation[39] came up with a plan that suggested that NASA "solidify pioneering" as the goal of NASA. NASA has truly lost its mission and its insight. It is not looking into the future. For NASA to look back into the future, it needs to feel secure in funding and support. That can only come from the White House, Congress and of course the people of the United States. NASA needs to know that it is appreciated and respected. It needs to "feel the love" in the colloquial sense. NASA can be brought back from the brink. The United States needs to bring it back. As we stand right now, the United States is sitting on a lonely launch pad looking at the stars. NASA needs to fight, NASA needs to revamp and most of all, NASA needs to believe in itself again. There is one other thing to think about…should NASA be privitized. This author believes it should not be. Privitizing NASA would be removing from the United States the level best of its space frontier achievements, and allowing commercialism to take over.

---

[38] "When I became the NASA administrator, (President Obama) charged me with three things," Bolden said in the interview which aired last week. "One, he wanted me to help re-inspire children to want to get into science and math; he wanted me to expand our international relationships; and third, and perhaps foremost, he wanted me to find a way to reach out to the Muslim world and engage much more with dominantly Muslim nations to help them feel good about their historic contribution to science, math and engineering." Quote from Space.com July 7, 2010

[39] The Space Foundation was founded in Colorado Springs in 1983, the Space Foundation is the world's premier organization to inspire, educate, connect, and advocate on behalf of the global space community.

Manned Space Flight and the CEV

On December 11, 2017 President Donald Trump signed a new directive setting the United States back on the path to the Moon and Mars officially. The Trump Directive's signing was attended by such Space/NASA special guests as Rockwell's Harrison Storms and moon landing veteran, Buzz Aldrin. Our current Vice President, Michael Pence was also in attendance as the head of the newly arisen National Space Council. In truth, after eight years of neglect with the Obama administration and his space policies or lack thereof, it felt good to see some sort of recognition for the space industry again. It also felt good to know that there is some acknowledgement of the amazing feats of robotic work that NASA has accomplished during the eight-year flat line policy of Obama and the 10 year sequestration that cut all services in the Military to the bone, not to mention NASA. While robotics were a great boon, what happened to the manned space program? Basically, it foundered. Yes, we sent astronauts back and forth to the ISS via the Russians and their aged Soyuz, paying heavily for the ride. What was accomplished, while all needed and necessary to any long term space mission, science testing, more long term space stays for the astronaut to figure out health and stamina but if there is nowhere to go, what was the use?

It's now time to look at just what NASA has to do to make the official mandate of returning to the moon and building a long term base that will set the United States up for a trip to Mars. There just isn't any other way to accomplish this. We need to solidify a base on the Moon that could support a long flight to Mars. The plan is to build the CEV, the new classification for Crew Exploration Vehicle.

The CEV

When the shuttle was retired July of 2011, NASA had already started a design for the next phase of manned spaceflight in 2004. That is the

Crew Exploration Vehicle. Along with that is the SLS or Space Launch System, which is the heavy lift rocket that will send the CEV on its way. On June 2005, the same contractors that had built the Apollo system and the Space shuttle met at the Walt Disney World Resort Hotel in Orlando, Florida. This "meeting" entailed exhibits set up by the contractors to show where they had been and where they were going to in the line of space efforts and to lobby for new business. The list included the "all stars" of the Aerospace world; Lockheed Martin, Northrop Grumman, and Boeing. Of course, there had been mergers between some of the greats like Lockheed and Martin Marietta, Northrop and Grumman, (Lunar module), while Boeing accumulated North American Rockwell (Space Shuttle) These giants had much under their very heavy belts as far as designing the next phase of American spacecraft. All these heavy hitters were not shy about showing their strength in the various Apollo, Space Habitat, International Space Station and more.

NASA already had their ideas about how they wanted this new space-ship to look. It was modeled right on the Apollo capsule of old. Perhaps it was a bit larger, able to hold four astronauts instead of two, but not much more than that. The person who is in charge of Orion or CEV or EM-1, all the same regardless of nomenclature, is Mike Hecker.[40] Hecker is a very realistic on how Orion will appear. As he said, "You can have a capsule or something with wings for coming back. All we told the industry is we need something that goes from the ground to low earth orbit and back." This led to the shape of the Apollo capsule, enlarged.

Boeing and Lockheed Martin led the two main design teams for com-petition to finally nail the NASA/CEV contest. Lockheed Martin wasted no time getting its design proposal in. The criteria of having to meet

---

[40] Capt. Michael Hecker, the agency's deputy administrator for development programs in the Office of Exploration for NASA.

something that could go into low earth orbit, like the shuttle, and be able to dock with the ISS, again like the shuttle, yet come up with something different was an enigma. If you had the shuttle, which did all of the above, why cancel it when it still had life left in it? Too much money? Did this exploration into designing the CEV really cost any less? Couldn't the shuttle have been used for the purpose it was designed for, space trucking and then design something for a lunar outpost or Mars flight? The shuttle was a proven entity. It did all of the above but because of the Columbia accident, fear took hold and it was canceled. The shuttle issues could have been dealt with, but no, we had to have a new spacecraft that did her job and be able to take care of a lunar flight or Mars flight as well. This reminds us of the F-111 TFX program that Secretary of Defense, Robert McNamara attempted to shove down all the Military services throats back in the 1960s. McNamara wanted the TFX to be all things to all services, much like the F-35 is today. The TFX didn't work, why should NASA follow the same beaten, broken path?

The plot grows thicker. NASA said that the designs cover the gamut, that's gamut as in range, length and scope, according the Mike Hecker. "We have ideas for a spacecraft that drops just like the Apollo capsule that plopped into the ocean after a mission. Or has wings for downrange flying. It's the gamut". In the early stages of the design, both wings and a blunt ended capsule were okay with NASA. However, the one thing all members of this design competition agreed on, was not to use any heat protection tiles, known in the Shuttle world as TPS or thermal protection system. As of right now, the PICA[41] and AV coat TPS systems are being looked at. Again, based on some of the old TPS technology from the shuttle. We know that the shuttle tiles were made to provide reusable protection from heat of re-entry. This also gave the Orbiter a sort of

---

[41] https://solarsystem.nasa.gov/docs/pr530.pdf RE TPS system for CEV

strange appearance if you looked at her close up. Some felt that she almost looked like she had been "tattooed". Basically, the reason for that was due to the serial numbers on the shuttle tiles that were corresponding to their place on the shuttle's surface. These tiles were all custom made to fit along the curvature of the Orbiter's body. Yet, the word had been spoken and the fix was in, no tiles on the CEV. It did make perfect sense to exclude tiles after the Columbia accident, no one wanted a repeat of that horror story.

Both Northrop Grumman and Lockheed Martin had serious ideas regarding carbon-carbon shield, which was a heavy reinforced material made notable with the Columbia disaster. Slate colored tiles were on the shuttle's belly. The lightweight white tiles were on the Orbiter's body. Because of a piece of foam from the external tank breaking loose during the launch, the Orbiter flew right into it damaging the leading edge of the left wing. This allowed the super –heated gases from re-entry to enter the aluminum frame of the shuttle and cause it to melt. Seven astronauts were killed. It did prove only one thing, the slate shield tiles held up better than the white light weight ones.

Safety was another factor in this design competition. There were so many of the former shuttle astronauts who complained long and hard that there was no escape system from the Orbiter. The question of launching humans on a solid rocket booster with solid rocket fuel was never a popular one. You cannot throttle back once those SRB motors are lit, like you can from a liquid fueled rocket.

Eileen Collins, the first female shuttle commander and a truly outspoken leader on crew escape systems, had excellent reasons for her position. While working with the Russians and their astronaut Vladimire Titov, a Soyuz capsule had a mission for the ISS. There was a launch pad fire and her flight mate Titov and the crew ejected safely from the emergency. They literally "walked away" from a potential disaster. It certainly did impress Collins!

In the early designs for the CEV, a launch escape system was devised for the rocket that would carry the CEV. It was on top of the CEV which was on top of the SLS. A small rocket would pull the CEV off and away from any potential disaster on the launch pad. The shuttle stack did not allow for that little luxury. After both the Challenger and Columbia disasters, a rather primitive system was devised allowing the astronauts to slip out on a pole, in their full pressure suits and parachutes and escape from the Orbiter but only after launch. It was a very limited system and truthfully not many held out any hope for it. The answer of the small rocket on top of the CEV capsule was going to be the safety mechanism.

Rockets

Another throw back to the Apollo days was the immensely powerful Saturn V rocket. The production line for this magnificent monster was shut down shortly after the Apollo construction line was closed down in 1973. The decision was made shortly after the Apollo 11 mission that they were not needed anymore. What a loss! The rocket was a spectacular piece of machinery that should have been held onto some way, but after using various sections of the huge rocket for other programs, it finally was left to history. Smaller boosters would be used to carry cargo up to the ISS or to launch satellites. NASA once again took a look at the past. The CONSTELLATION program which encompassed the ARES 1 and ARES 5 rockets was put into place by President George W. Bush on January14, 2004. The Shuttle would be retired in 2010 and the Constellation program would be set to take over handling the ISS and start moving towards the hopes of a lunar base. Yet, with all the hopes, NASA dropped the ball because it could not come up with a solid program to do all of this. When President Obama came into office, one of the first things he did was shut down the Constellation program stating that it was too expensive and had no valid direction. The country didn't need it and it was cut from the budget in 2010. In many ways, the Obama administration really had no use for NASA and left it to founder, looking for a program and a hope of someone caring about its mission. That was something that Obama had no use for. The Obama Administration decided to create the Augustine Panel to solidify his wants for all of NASA. Obama and his "Flexible Plan for Mars" was seriously nothing more than a place holder for something he didn't wish to have his administration bothered with. NASA sat, its budget cut to the literal bone, using robotics to complete some programs that were already in the pipeline, also sending astronauts to the ISS via the Russians at $70 million a trip.

This was money that NASA could have used to work on other projects. NASA also relied on the hopes of commercial space ventures. People like Elon Musk and his Space X company along with others in the private sector started to develop some cargo rockets and maybe a dream of a human rated spacecraft. While it did take time, truly in a relatively short amount of time, Musk had come up with the little Dragon spacecraft that really saved the day in bringing cargo back and forth to the ISS. While it is designed to carry humans, it is not yet human rated. However, thanks to the entrepreneurship of people like Musk, the U.S. kept its hand is the space program.

Another look at the CEV

The Disney World Conference of September 19, 2005 also included Mike Griffin, the new NASA administrator as of April, 2005. Griffin was an aerospace engineer who really understood what was going and the problems that NASA faced. He was known for being very straightforward and not always politically correct. It was just what NASA needed. Griffin had a true idea of what was needed.

He rocked the Conference and the contractors by proposing expendable rockets like the Delta 4 and Atlas 5. Griffin had his own ideas on the revamped lunar program. He saw it this way:

a. First return mission to the moon --lifting off on two rockets not just one. This very thin, needle like rocket would place the crew on top and that would fly only after a larger cargo rocket (built on shuttle technology) was launched. The cargo rocket would carry the prize package, the Lunar Lander.

b. The two ship formation would dock and head to the moon, much like the Apollo missions of old. It would take a week to get to the moon in this fashion.

According to Griffin, "Call it Apollo on steroids!" There would be four astronauts that would ride to the moon, compared to the three that were in the Apollo mission. A week long trip to the moon would take the crown from Apollo's three astronauts at half the cost, so says NASA's number crunchers. Griffin felt the program would make it without the extra money from Congress, costing an estimated $104 billion. If you look at this, it's sort of a lonely and sad projection and it also looks like someone was reading from the USSR play book. An early Soviet plan to beat the U.S. to the moon had their ideas of launching parts of the Soyuz moon ship, one rocket at a time. It also built on the old designs instead of exploring a new one. The current Soyuz vehicle can trace its heritage back to the first Soyuz that flew back in 1967, except it had a horrific accident. The very first Soyuz had an electronics panel fail which resulted in the parachute lines tangling during re-entry. The Soyuz is a land recovery, unlike the Apollo, which is a water recovery. Due to the tangled parachute, the Soyuz hit the ground, killing the astronaut on board. This machine is still carrying U.S. astronauts to the ISS and still making rather rattling, nerve wracking land recoveries.

Mike Griffin's moon perspective which was only partial, still had some good things in it:

a. Sparing the NASA crews a hard transition as the nation traded the shuttle for this not so new method of reaching the Moon and Mars.

b. Official estimates from Kennedy Space Center showed the work force that maintained the shuttle, would fall following the retirement of the shuttle. This would once again, decimate the surrounding communities where these workers lived for so many years.

c. Proponents of the shuttle derived designs believed that the more jobs that might have been salvaged as the U.S. moved from the shuttle to the CEV.

Without a doubt, the thought of many Floridian constituents losing jobs and homes did   make Congress and the Senate members feeling pretty queasy about it. Many did not want to sign onto Griffin's plans. The glory days of Apollo in Broward County, Florida were also remembered for the most negative impact it created as Apollo phased out. NASA's own workers held onto their jobs, while the contractors lost theirs as the shuttle transitioned in during 1981. The contract workers lost their jobs in the thousands and were forced to sell homes, uproot kids and move away, from schools causing turmoil in the area. When the word that the shuttle was going to retire was heard, it also brought all sorts of nightmares of the post Apollo depression era back to Broward county, again.

On August 3, 2006, NASA announced Lockheed as the new contractor of the CEV. NASA was planning to announce the new contract from the ISS and arranged for astronaut, Jeff Williams to do a video from the ISS to help unveil the new contractor and selection of the spacecraft's name. We've been calling it Crew Exploration Vehicle for several years."

Williams explained, "But today it has a name.... ORION." The message was heard by the press over the air to ground radio on August 22, 2006, not at the end of the month when NASA intended to make it public. Headlines had ORION's new name plastered all over the place. On August 31,2006 with the announcement made, Lockheed had the contracts, albeit without the expected fanfare. Lockheed still had the $8 million for the early phases of the spacecraft. It also meant a lot of work between the contractors and astronauts of NASA.

More new Rockets

Regardless of the snafu of the CEV naming, NASA actually had two names for the new capsule. ORION and ARES were two distinct names and types of capsule and rocket which would send both people and cargo into a low earth orbit on missions to the ISS, and later the moon and Mars. They are both direct descendants of the space shuttle technology. An early artist's rendering for the unmanned test flights brought it home.

However, NASA still needed a number of launches to confirm the soundness of the ARES design before allowing it to carry humans. As it came to pass, one flight utilized hardware left over from the first shuttle lift off since the 2003 Columbia disaster, not exactly auspicious.

On the Discovery STS-114 mission, there were two sets of SRBs. One pair was replaced before liftoff by a second set. Four cylindrical segments of one of the discarded boosters were selected for test use in the ARES I. This would be the rocket to send the astronauts inside the CEV capsule, into low earth orbit. There were similar tests planned for the ARES 5 cargo rocket that would carry the lunar-lander for moon landing. However, ARES was not long lived. ARES went under the CONSTELLATION Program that was proposed by President George W. Bush, to send the U.S. back to the moon and beyond. Bush proposed CONSTELLATION as he was shutting the shuttle down, way too early for this author's likes. It was really short-lived. As soon as President Obama came into office, CONSTELLATION was canceled quickly with Obama stating that it was an ineffectual program that harbored on no new ideas and too much money for nothing. It was off the budget by 2011, leaving NASA to once again founder and wonder what in hell it really was doing here if it was too much money to keep it going with space as its prime directive.

Escape system...PLEASE!!!!

For all of this, one problem had not been solved, that problem is a major issue among the astronauts; a quick escape system. They wanted to make sure that their voices were heard. The Astronauts Corp. for NASA was busy lobbying for a crew escape system before the White House announcement that the shuttle would be retired. The astronauts were also invited to take part in designing the inside of the ORION/CEV capsule. One of the astronauts and veteran space walker, Lee Morin, would use his experience in design with interacting cockpit equipment for fighter jets. Morin was also the deputy manager of the Astronaut office and the ORION/CEV cockpit task group manager.

It was already a working fact that the room in Orion/CEV, which the astronauts had to move around in was terribly short in comparison to the space shuttle. When you consider the astronauts were used to the two story shuttle crew compartment that had pilots on the upper deck and the mid deck held the place for eating, sleeping and experiments, also a little privacy for personal needs, aka a toilet. Cramped is a good way to describe the four astronauts who would be traveling in the CEV.

By contrast, astronauts arriving at the ISS and aboard the CEV might be motivated to get out of the ORION/CEV and into the ISS just to be able to stretch their legs, literally. This leaves us to consider the long week it would take to reach the moon, not to mention the long weeks it would take to reach Mars. Stress from being in such a tight space for such a long time will play on the astronauts and it's a big factor. Astronauts, Norm Thagard, Mike Finke and Ed Lee made their trips to the ISS via the Soyuz capsule courtesy of the Russians. They were strapped in with their knees in their chests, dressed in bulky suits and had their duffel bags stowed inches from their heads. ISS must have seemed like a luxury hotel when they arrived! "You use all the nooks and crannies in a Soyuz", said Lee Morin. "You see things like clothing you need for survival situations stuffed everywhere!"

Beyond the configuration of the Orion/CEV spacecraft, NASA will also have to deal with spacesuits the CEV crew will have to wear. Remember the shuttle suits were designed not for space walks but for the bail out survival option to be used (aka: the long pole to slip out of the shuttle) shortly after liftoff or just before landing. Those 90lbs suits have a survival raft for water landings and a dive marker for signals along with other survival gear. NASA will need to come up with suits light and strong enough to deal with the environment in the CEV or lack of it. The CEV will have no more than four windows where the shuttle had eleven. While the CEV will be much less complicated than the shuttle interior, it will have one central control panel with a few touch screen computers which will surround large buttons for crew use. NASA did want one thing and that was a "paperless" cabin.

The astronauts were not sitting still for anything. They wanted their say in every aspect of CEV most especially the facilities that pertained to their everyday necessities like the toilet. The windows were a major issue. Members of the astronaut office took the early blueprints of the capsule layout to a special room called the "Reconfigurable Orbital Cockpit Facility". This is the room where computers were used to cut Styrofoam mockups of the window frames for judgment of the field of vision.

The CEV will also need a lot less piloting than the shuttle. The CEV will land like Apollo, with re-entry into the water via parachute, which is making the CEV pilots not terribly pleased. If you remember way back in the Mercury program, piloting was a serious issue. The spacecraft that was designed caused a head-on war between the astronauts and the designers. The Astronauts were not going to be "spam in a can". They demanded the right to pilot their craft in case of any emergency. The Mercury Seven got that "wish" after basically refusing to fly.

The pros and cons of going with the "Apollo style" spacecraft

As of May 18, 2005, with members of the U.S. Senate Sub Committee for Science and Space, which had concerns about whether the new Moon and Mars program could be viable long term. "I do not want", said Mike Griffin, "…we do not want to repeat the mistake of the Apollo program, where money caused unique capabilities to be shut down directly and irretrievably."

There were some tough lessons learned from the Apollo and Shuttle programs. Both a new spacecraft and mission is important but more urgent than that is keeping the astronauts alive. That will depend not only on engineering, it will depend on how generous the sitting Congress is about what they will pay to remain the leader in the world's contenders for the title of first in space. Will Congress be willing to pay that hefty price tag to do the mission properly? The price could be in the billions. Or, will Congress be satisfied for a piece meal situation, much like the cheap, catch as catch can method with which the shuttle was involved.

NASA already has been accused of being an agency with only one thing on its mind and that is a lunar base and a trip to Mars. That is not true. NASA delves into many things like robotic space travel for one. The CASSINI mission was the first truly in depth study of Saturn, its moons and the rings, which were a total mystery till now, ending the flight with a spectacular burn up in Saturn's atmosphere. That wasn't the only phenomenal mission with robots. There is also the Mars Rovers, Spirit and Opportunity who had been turning out information about the Martian surface for six years now and the amount of data is staggering. NASA cannot be accused of tunnel vision, not when its turning out programs like this. What is lacking for NASA is a true commitment to manned space flight. Back in 1992-98, Dr. John Gibbons who was the science advisor for President Bill Clinton, was a proponent of the robotic mission. He and a small group of scientists who believed that spending large amounts for money on sending astronauts on manned missions was basically stupid. In his quote "And I think anyone who knows anything about astronomy would say that's mad!" referring to the manned missions. Gibbons really felt that the robotic missions did a better job of exploring than the astronauts did. There was a true unwillingness to fund the CEV. Gibbons also hated every aspect of the CONSTELLATION program. This was a sharp contrast to all the fighting to save funding for the ISS. The ISS program stalled till Gibbons and Clinton decided to advocate building the complex. The ISS station was seen as the way for nations around the world to work together on non- military projects. That was a lovely thought but the money from foreign nations was slow to come by, especially Russia who actually ended up with Dan Goldin, then NASA administrator who paid off the Russian debt to the ISS with the money from the Shuttle's safety upgrade funding. Programs that were needed to upgrade the shuttle were kicked to the curb. In Gibbons' thoughts, "I believe going to Mars but we should go robotically to send our senses there, but not our bodies. This sending people to Mars stuff, is a dreamlike plan to escape the realities of earth." However, you do have those that feel it's much sexier to have your photo taken with a real live astronaut that some piece of metal space probe.

Epilogue

The Next Step

All of those astronauts that did fly with the shuttle knew that once the shuttle retired, their flight time was going to be cut and quickly. Their days at NASA were most likely over as there were not enough slots going to the ISS to cover all of them. From the two hundred astronauts that NASA had at the end of the shuttle program, the list was cut down to possibly fifty. That is a big cut. It's a big cut in training, money spent in training, time on behalf of the astronauts, some of who never got to fly and basically got tired waiting.

Retiring the shuttle was also a chance for NASA to say adieu to some really sad, painful memories. Not only the two shuttles and fourteen astronauts lost, but the lack of mission. The shuttle never really had a mission to call her own, other than being a space truck. Sadly, the shuttle's history shows a true lack of focus as far as NASA was concerned and the need to make compromise after compromise to keep the program solvent. The ISS was really the only thing that the shuttle had. Yes, she did do some very hazardous spacewalks, repairs of the Hubble telescope, help to the ISS when needed along with delivering the construction pieces for the ISS when building. However, after that what really was there? When the ISS was completed, there was no more work to do there, so it was back to shuffling astronauts back and forth and doing endless runs for supplies and dropping and returning astronauts off at the ISS and launching satellites for the USAF.

NASA also did try to use its customer, the USAF as a means to an end. The Shuttle was to supply the USAF with a cheap and reliable way to deliver payloads to a low earth orbit. NASA was also hoping to garner more funds for the shuttle by using the USAF as a customer and partner when it went to Congress for money. There was thirty percent of the payloads that the Shuttle carried were from the USAF. The USAF moved to retire the fleet of Titan rockets is had been using to launch its satellites. The plan was to cover all the political bases, but it wasn't helpful to those who were doing the negotiating. According the Roberts Seamans Jr, Secretary of the USAF, "This is kind of shocking. Why put astronauts at risk to put unmanned satellites in space?" It was a valid question, but anything to keep the shuttle running. Seamans also felt that it was "asinine to launch every payload on the shuttle risking the five astronauts per flight." Yet, the USAF advisors thought the idea was worth looking at and NASA did seem enthusiastic enough to want to suit the shuttle to meet the military needs and requirements. The laundry list for the USAF's version of the shuttle was distinct enough. Again, Robert Seamans added that the dumping of the external tank and the SRBs after a launch would be equivalent to dumping a 747 aircraft after every trip across the Atlantic Ocean. With thoughts like this floating around it was no wonder there was such a shadow posted over the most complex and successful spacecraft ever flown. Don't doubt it. The shuttle for all of her problems was one of the most fantastic programs NASA ever ran, next to Apollo.

As we look at the CEV program and what is going on around it, we once again see part of NASA's on going problems, first money, second a mission that is viable and third, the wherewithal to pull the program together. The design phase is still on going, while the spacecraft is being built. A slow process of what should NASA do next and will it work, prevails politically. Thankfully for NASA, President Trump has signed the new directive for NASA to move towards the Moon and Mars. However, does that give the program the heart and guts to do what needs to be done. There is much to do. The SLS still needs to be tested, while the CEV is being built, tested and streamlined for service. The question needs to seriously be asked, is this what we really need to reach Mars? Will an established lunar base be able to support the program to Mars? Is the CEV a cheap way out? Following the dictates of the successful Apollo, will it be enough to reach Mars? Four astronauts tucked away is a very small capsule, travelling for weeks through deep space to get to another planet, will the CEV truly be what is needed or is NASA settling for what they can get instead of what they should have. It's a very serious question. To be sure, has it really been seriously talked about and decided on. Many are yet to be convinced and many are not convinced. The argument still rages on that robotics are enough, why risk lives? It's a good argument. However, what can NASA show that will prove to those naysayers that manned space flight is the next move. Man needs to branch out and see what is in the near galaxies, the nearest planets like Mars. NASA will have to do a better job of proving its concept and Congress has to be more open-minded in dealing with the concept of manned space flight. Congress must feel it is a positive move to give NASA what it needs to get the job done, not on the cheap but with all it needs, just like it did with Apollo. Many will say that Apollo got all is needed because of the assassination of a beloved president and a mandate that he had given NASA before he died. It was a powerful reason to push ahead to the moon. What does NASA have now to show it is necessary to build a lunar base and move on to Mars. NASA needs to find that key if it is going to make the CEV, the SLS and its mission a viable and seriously planned undertaking with all the things it needs to do it and not begging for crumbs off the table of Congress.

NASA does need a shot in the arm to get it back to where it could operate internally, externally and politically without a lot of interference. NASA, has done some amazing things on the tiny budget that it has. Yet, if it wishes to reach out to the stars and planets in manned missions once again, NASA needs a solid commitment from Congress, and the country itself, if it is going to complete this very elusive undertaking. Most importantly, NASA needs the support of the United States behind it, just like it had with Apollo. Given the light of the international stage, it may be a hard order to come up with. The country right now is in the throes of political strife. Democrats and Republicans are at each other throats and it seems the last thing on anyone's mind is should NASA go to the Moon. While we strive to get our kids back on the path of a good, solid education that supports math and science, they need something to apply that math and science to. That should come from NASA, it should be exploited by NASA. Our kids need something to strive for.

Back in 1964-65, there was a little show in Flushing, New York called the 1964-65 World's Fair. That World's Fair came during the hottest part of the Cold War, things had heated up with Russia, the United States watched its beloved, young president murdered in the streets of Dallas, Texas. The country was unsure, in pain, and sad. There were many issues with racial integration, Even at this World's Fair, there was picketing by the CORE[42] because of the Fair's hiring practices. However, there is one thing this Fair did have, hope for the future. It had a view of what the future might look like with programs like General Motor's Futurama. As a kid, this author watched as the lunar rovers crawled over the surface of the moon, saw underwater houses, and cities and highways that were fast and clean. The Hall of Science, which is now a beautiful Science museum dedicated to kids, showed the docking of what would be the Apollo spacecraft and the lunar lander in the 1960s. Outside, there were rockets!! A section of the Saturn V, the Atlas and the lunar lander mock up which you could walk up and touch. There was also the X-15 rocket plane mounted on a pedestal. It was really awe inspiring. It led to

---

[42] Congress on Racial Equality

this author a career in aviation. Today, we have many of the same issues, albeit different levels of the same issues that we had back in the 1960s.

Today, we face a world of cellphones and self absorption. We face our kids who have had their educational processes weakened in schools that are overcrowded, over sensitized, and political cesspools instead of a place where a kid could learn to dream. That and the fact that we are too busy worrying about what is going on in Facebook, instead of what is going on at our very lonely dinner tables at night. Lonely because these dinner tables are no longer used as a place for a family to come together and talk. How does space come into this? We have so much technology, we have become desensitized to the prospect of going back to the Moon or making that trip to Mars. It's old thinking according to some. Send a robot, why send a bunch of astronauts?

Do we even need to go out into space? The answer is yes, we do. If we want to expand the horizons of this very small planet we live on, if we want to learn and explore, if we want to go "where no man has gone before", we NEED NASA and we NEED to support our space program. We will be damning this world to its own demise, and quickly, if we lose that exploration will. As to our government, we need to let them know this is important to us as citizens, work to show our kids there is some-thing more than a cellphone that you use to text your buddy who is sitting across from you, instead of talking to that person. We need to see the stars and ask why and how did they get that way. Science needs to be made a reality again. NASA needs to survive. The CEV is a chance at that and Congress, the White House, and our own people need to find that magic again. NASA needs to find "failure is not an option" and make that magic happen again. For all those that we have lost to the trials of spaceflight and space travel, we must succeed. For all that has been learned, we must succeed. NASA and this country can't allow the demise of the United States manned spaceflight program.

Appendices

A.

National Space Act of 1958

A. The expansion of human knowledge in both the atmosphere and space.

B. It was intended to foster improvements in performance, speed, safety in the proficiency of space vehicles and aircraft.

C. The Act defines NASA's path. The development and operation of vehicles that are capable of carrying instruments, equipment and supplies, and/or living organisms through space and return.

D. It cites the need to conduct long-range studies to discover potential benefits and opportunities and problems in both aeronautical and space activities.

E. It challenges NASA to maintain the United States as a leader in the space science and technology.

F. It also provides for relevant information to be passed on to the Department of Defense and related agencies.

G. The Act directs NASA to facilitate the cooperation of nations within the peaceful pursuit of aeronautical and space activities.

H. It directs the cooperation and coordination with other public agencies to avoid "unnecessary duplication of effort, facilities and equipments."

The Functions of NASA

* *To plan and direct and conduct aeronautical and space activities.*

* *Arrange for participation by the scientific community in planning scientific measurements and observations to be made through use of*

aeronautical and space vehicles, and conduct or arrange for the conduct of such measurements and observations and:

* Provide for the widest practicable and appropriate dissemination of information concerning its activities and the results thereof.

* In the performance of its function the Administration is authorized:

* To make, promulgate, issue, rescind, and amend rules and regulations governing the manner of its operations and the exercise of the powers vested in it by law;

To appoint and fix the compensation of such officers and employees as may be necessary to carry out such functions. Such officers and employees shall be appointed in accordance with the civil service law and their compensation fixed in accordance with the Classification Act of 1949, except that

1. There is a maximum of 10 positions of no more than two hundred and sixty of the scientific, engineering and administrative personnel of the Administration without regard to such laws,

2. To the extent of the administrator, deems such action necessary to recruit specially qualified scientific and engineering talent, he may establish the entrance grade for scientific and engineering personnel without previous service in the Federal Government at a level up to two grades higher that the grade preceded for such personnel, under the General Schedule established by the Classification Act of 1949, and fix their compensation accordingly;

B.

The Functions of NASA

* To plan and direct and conduct aeronautical and space activities.

* Arrange for participation by the scientific community in planning scientific measurements and observations to be made through use of

aeronautical and space vehicles, and conduct or arrange for the conduct of such measurements and observations and:

* Provide for the widest practicable and appropriate dissemination of information concerning its activities and the results thereof.

*. In the performance of its function the Administration is authorized:

* To make, promulgate, issue, rescind, and amend rules and regulations governing the manner of its operations and the exercise of the powers vested in it by law; To appoint and fix the compensation of such officers and employees as may be necessary to carry out such functions. Such officers and employees shall be appointed in accordance with the civil service law and their compensation fixed in accordance with the Classification Act of 1949, except that

1. There is a maximum of 10 positions of no more than two hundred and sixty of the scientific, engineering and administrative personnel of the Administration without regard to such laws,

2. To the extent of the administrator, deems such action necessary to recruit specially qualified scientific and engineering talent, he may establish the entrance grade for scientific and engineering personnel without previous service in the Federal Government at a level up to two grades higher that the grade preceded for such personnel, under the General Schedule established by the Classification Act of 1949, and fix their compensation accordingly; To acquire (by purchase, lease, condemnation or otherwise) construct, improve, repair, and maintain laboratories, research and testing sites and facilities, aeronautical and space vehicle quarters and related accommodations for employees and dependents of employees of the Administration and other real or personal property (including patents) or any interest therein, deems necessary within and outside the continental United States; to lease to other such real and personal property; to sell and otherwise dispose of real and personal property (including patents and rights) in accordance with the provisions of the Federal Property and Administrative Service Act of 1949 and to provide by contract or otherwise for cafeterias and other necessary facilities for the welfare of employees of the Administration at its installations and purchase and maintain equipment.

1. To accept unconditional gifts or donations of services, money, or property, real personal or mixed tangible or intangible;

Without regard to section 3648 of the revised statutes, as amended to enter into and perform such contract, leases, cooperative agreements, or other transactions as may be necessary in the conduct of its work and on such terms as it may deem appropriate, with any agency or instrumentality, of the United States, or with any State, territory, or possession, or with any political subdivision, or with any person, firm, association, corporation, education institutions to the maximum extent practicable and consistent with the accomplishment of the purpose of this Act, such contracts, leases agreements and other transactions that shall be allocated by the Administrator in a manner which will enable small business concerns to participate equitably and proportionately in the conduct of the work of the Administration.

2. To use with consent, the services equipment, personnel and facilities of Federal and other agencies with or without reimbursement, and on a similar basis to cooperate with other public and private agencies and instrumentalities in the use of the services, equipment and facilities available to the Administration and any such department or agency is authorized, notwithstanding any other provision of law to transfer or to receive from the Administration, without reimbursement, aeronautical and space vehicles and supplies and equipment other than administrative supplies and equipment.

Further more:

A. To appoint advisory committees as may be appropriate for purposes of consultation and advice to the Administration in the performance of its functions:

B. To establish within the administration such offices and procedures as may be appropriate to provide for the greatest possible coordination of its activities under this Act with related scientific and other activities being carried on by other public and private agencies and organizations.

C. When determined by the administrator to be necessary, and subject to such security investigations as he may determine to be appropriate, to employ aliens without regard to statutory provisions prohibiting payment of compensation to aliens. To employ retired commissioned officers of the armed forces of the United States and compensate them at the established rate for the positions occupied by them with in the Administration, subject only to the limitations in pay set forth in section 212 of the Act of June 30, 1932.

With the approval of the President, to enter into cooperative agreements under which members or the Army, Navy, Air Force, and Marine Corps may be detailed by the appropriate Secretary for the services in the performance of functions under this Act to the same extent as that to which they might be lawfully assigned in the Department of Defense.

D. To consider, ascertain, adjust, determine, settle and pay on behalf of the United States in full satisfaction thereof, any claim for $5000 or less against the United States, for bodily injury, death or damage to or loss of real or personal property resulting from the conduct of the Administrations functions as specified in subsection (a) of this section, where such claim is presented to the Administration in writing within two years after the accident or incident out of which the claim arises.

C.

Findings from Chapter 6 of the Columbia accident investigation board Findings

F6.1−1 NASA has not followed its own rules and requirements: on foam-shedding. Although the agency continuously worked on the foam-shedding problem, the debris impact requirements have not been met on any mission.

F6.1−2 Foam-shedding, which had initially raised serious safety concerns, evolved into "in-family" or "no safety-of-light" events or were deemed an "accepted risk."

F6.1−3 Five of the seven bipod ramp events occurred on missions flown by Columbia, a seemingly high number. This observation is likely due to Columbia having been equipped with umbilical cameras earlier than other Orbiters.

F6.1−4 There is lack of effective processes for feedback or integration among project elements in the resolution of In-Flight Anomalies.

F6.1−5 Foam bipod debris-shedding incidents on STS-52 and STS-62 were undetected at the time they occurred, and were not discovered until the Board directed NASA to examine External Tank separation images more closely.

F6.1−6 Foam bipod debris-shedding events were classified as In-Flight Anomalies up until STS-112, which was the first known bipod foam-shedding event not classified as an In-Flight Anomaly.

F6.1−7 The STS-112 assignment for the External Tank Project to "identify the cause and corrective action of the bipod ramp foam loss event" was not due until after the planned launch of STS-113, and then slipped to after the launch of STS-107.

F6.1−8 No External Tank configuration changes were made after the bipod foam loss on STS-112.

F6.1–9 Although it is sometimes possible to obtain imagery of night launches because of light provided by the Solid Rocket Motor plume, no imagery was obtained for STS-113.

F6.1–10 NASA failed to adequately perform trend analysis on foam losses. This greatly hampered the agency's ability to make informed decisions about foam losses.

F6.1–11 Despite the constant shedding of foam, the Shuttle Program did little to harden the Orbiter against foam impacts through upgrades to the Thermal Protection System. Without impact resistance and strength requirements that are calibrated to the energy of debris likely to impact the Orbiter, certification of new Thermal Protection System tile will not adequately address the threat posed by debris.

Recommendations:

• None

6.2 SCHEDULE PRESSURE

D.

CAIB

Email from Linda Ham to Ron Dittenmore regarding ET briefings:

-----Original Message----- From: HAM, LINDA J. (JSC-MA2) (NASA)

Sent: Wednesday, January 22, 2003 10:16 AM

To: DITTEMORE, RONALD D. (JSC-MA) (NASA)

Subject: RE: ET Briefing - STS-112 Foam Loss Yes, I remember....It was not good. I told Jerry to address it at the ORR next Tuesday (even though he won't have any more data and it really doesn't impact Orbiter roll to the VAB). I just want him to be thinking hard about this now, not wait until IFA review to get a formal action. [ORR=Orbiter Rollout Review, VAB=Vehicle Assembly Building, IFA=In-Flight Anomaly]

E.

CAIB Volume 1: DECISION-MAKING DURING THE FLIGHT OF STS-107 Initial Foam Strike Identification:

As soon as Columbia reached orbit on the morning of January 16, 2003, NASA's Intercenter Photo Working Group began reviewing liftoff imagery by video and film cameras on the launch pad and at other sites at and nearby the Kennedy Space Center.

The debris strike was not seen during the first review of video imagery by tracking cameras, but it was noticed at 9:30 a.m. EST the next day,

Flight Day Two, by Intercenter Photo Working Group engineers at Marshall Space Flight Center.

Within an hour, Intercenter Photo Working Group personnel at Kennedy also identified the strike on higher-resolution film images that had just been developed. The images revealed that a large piece of debris from the left bipod area of the External Tank had struck the Orbiter's left wing. Because the resulting shower of post-impact fragments could not be seen passing over the top of the wing, analysts concluded that the debris had apparently impacted the left wing below the leading edge.

Intercenter Photo Working Group members were concerned about the size of the object and the apparent momentum of the strike. In searching for better views, Intercenter Photo Working Group members realized that none of the other cameras provided a higher-quality view of the impact and the potential damage to the Orbiter. Of the dozen ground-based camera sites used to obtain images of the ascent for engineering analyses, each of which has film and video cameras, five are designed to track the Shuttle from liftoff until it is out of view. Due to expected angle of view and atmospheric limitations, two sites did not capture the debris event.

Of the remaining three sites positioned to "see" at least a portion of the event, none provided a clear view of the actual debris impact to the wing. The first site lost track of Columbia on ascent, the second site was out of focus – because of an improperly maintained lens – and the third site captured only a view of the upper side of Columbia's left wing. The

Board notes that camera problems also hindered the Challenger investigation.

Over the years, it appears that due to budget and camera-team staff cuts, NASA's ability to track ascending Shuttles has atrophied – a development that reflects NASA's disregard of the developmental nature of the Shuttle's technology. (See recommendation R3.4-1.) Because they had no sufficiently resolved pictures with which to determine potential damage, and having never seen such a large piece of debris strike the Orbiter so late in ascent, Intercenter Photo Working Group members decided to ask for ground-based imagery of Columbia.

IMAGERY REQUEST 1 To accomplish this, the Intercenter Photo Working Groups Chair, Bob Page, contacted Wayne Hale, the Shuttle Program Manager for Launch Integration at Kennedy Space Center, to request imagery of Columbia's left wing on-orbit. Hale, who agreed to explore the possibility, holds a Top Secret clearance and was familiar with the process for requesting military imaging from his experience as a Mission Control Flight Director. This would be the first of three discrete requests for imagery by a NASA engineer or manager. In addition to these three requests, there were, by the Boards count, at least eight "missed opportunities" where actions may have resulted in the discovery of debris damage.

Shortly after confirming the debris hit, Intercenter Photo Working Group members distributed a "L+1" (Launch plus one day) report and digitized clips of the strike via e-mail throughout the NASA and contractor communities. This report provided an initial view of the foam strike and served as the basis for subsequent decisions and actions. Mission Managements Response to the Foam Strike.

As soon as the Intercenter Working Group report was distributed, engineers and technical managers from NASA, United Space Alliance, and Boeing began responding. Engineers and managers from Kennedy

Space Center called engineers and Program managers at Johnson Space Center. United Space Alliance and Boeing employees exchanged e-mails with details of the initial film analysis and the work in progress to determine the result of the impact. Details of the strike, actions taken in response to the impact, and records of telephone conversations were documented in the Mission Control operational log. The following section recounts in chronological order many of these exchanges and provides insight into why, in spite of the debris strike severity, NASA managers ultimately declined to request images of Columbia's left wing on-orbit.

Flight Day Two, Friday, January 17, 2003 In the Mission Evaluation Room, a support function of the Shuttle Program office that supplies engineering expertise for missions in progress, a set of consoles are staffed by engineers and technical managers from NASA and contractor organizations. For record keeping, each Mission Evaluation Room member types mission-related comments into a running log. A log entry by a Mission Evaluation Room manager at 10:58 a.m. Central Standard Time noted that the vehicle may have sustained damage from a debris strike.

"John Disler [a photo lab engineer at Johnson Space Center] called to report a debris hit on the vehicle. The debris appears to originate from the ET Forward Bipod area...travels down the left side and hits the left wing leading edge near the fuselage...The launch video review team at KSC think that the vehicle may have been damaged by the impact. Bill Reeves and Mike Stoner (USA SAM) were notified." [ET=External Tank, KSC=Kennedy Space Center, USA SAM=United Space Alliance Subsystem Area Manager]

At 3:15 p.m., Bob Page, Chair of the Intercenter Photo Working Group, contacted Wayne Hale, the Shuttle Program Manager for Launch Integration at Kennedy Space Center, and Lambert Austin, the head of the Space Shuttle Systems Integration at Johnson Space Center, to inform them that Boeing was performing an analysis to determine trajectories, velocities, angles, and energies for the debris impact. Page also stated that photo-analysis would continue over the Martin Luther King Jr. holiday weekend as additional film from tracking cameras was developed. Shortly thereafter, Wayne Hale telephoned Linda Ham, Chair of the Mission Management Team, and Ron Dittenmore, Space Shuttle Program Manager, to pass along information about the debris strike and let them know that a formal report would be issued by the end of the day.

John Disler, a member of the Intercenter Photo Working Group, notified the Mission Evaluation Room manager that a newly formed group of analysts, to be known as the Debris Assessment Team, needed the entire weekend to conduct a more thorough analysis. Meanwhile, early opinions about Reinforced Carbon-Carbon (RCC) resiliency were circulated via e-mail between United Space Alliance technical managers and NASA engineers, which may have contributed to a mindset that foam hitting the RCC was not a concern. -----

EMAIL
Original Message-----
From: Stoner-1, Michael D
Sent: Friday, January 17, 2003 4:03 PM
To: Woodworth, Warren H; Reeves, William D Cc: Wilder, James; White, Doug; Bitner, Barbara K; Blank, Donald E; Cooper, Curt W; Gordon, Michael P.

Subject: RE: STS 107 Debris Just spoke with Calvin and Mike Gordon (RCC SSM) about the impact. Basically the RCC is extremely resilient to impact type damage. The piece of debris (most likely foam/ice) looked like it most likely impacted the WLE RCC and broke apart. It didn't look like a big enough piece to pose any serious threat to the system and Mike Gordon the RCC SSM concurs. At T +81seconds the piece wouldn't have had enough energy to create a large damage to the RCC WLE system. Plus they have analysis that says they have a single mission safe re-entry in case of impact that penetrates the system. As far as the tile go in the wing leading edge area they are thicker than required (taper in the outer mold line) and can handle a large area of shallow damage which is what this event most likely would have caused. They have impact data that says the structure would get slightly hotter but still be OK. Mike Stoner USA TPS SAM [RCC=Reinforced Carbon-Carbon, SSM=Sub-system Manager, WLE=Wing Leading Edge, TTPS=Thermal Protection System, SAM= Sub-system Area Manager]

F.

Transcript Excerpts from the January 21, Mission Management Team Meeting.

CAIB Volume 1

Ham: "Alright, I know you guys are looking at the debris."

McCormack: "Yeah, as everybody knows, we took a hit on the, somewhere on the left wing leading edge and the photo TV guys have completed I think, pretty much their work although I'm sure they are reviewing their stuff and they've given us an approximate size for the debris and approximate area for where it came from and approximately where it hit, so we are talking about doing some sort of parametric type of analysis and also we're talking about what you can do in the event we have some damage there."

Ham: "That comment, I was thinking that the light rationale at the FRR from tank and orbiter from STS-112 was.... I'm not sure that the area is exactly the same where the foam came from but the carrier properties and density of the foam wouldn't do any damage. So we ought to pull that along with the 87 data where we had some damage, pull this data from 112 or whatever light it was and make sure that…you know I hope that we had good light rationale then."

McCormack: "Yeah, and we'll look at that, you mentioned 87, you know we saw some fairly significant damage in the area between RCC panels 8 and 9 and the main landing gear door on the bottom on STS-87 we did some analysis prior to STS-89 so uh…"

347

Ham: "And I'm really I don't think there is much we can do so it's not really a factor during the light because there is not much we can do about it. But what I'm really interested in is making sure our light rationale to go was good, and maybe this is foam from a different area and I'm not sure and it may not be co-related, but you can try to see what we have."

McCormack: "Okay."

After the meeting, the rationale for continuing to fly after the STS-112 foam loss was sent to Ham for review. She then exchanged e-mails with her boss, Space Shuttle Program Manager Ron Dittenmore:

---Original Message-----
From: DITTEMORE, RONALD D. (JSC-MA) (NASA)
 Sent: Wednesday, January 22, 2003 9:14 AM To: HAM, LINDA J. (JSC-MA2) (NASA)
Subject: RE: ET Briefing - STS-112 Foam Loss
You remember the briefing! Jerry did it and had to go out and say that the hazard report had not changed and that the risk had not changed...But it is worth looking at again.

G.

From Decision Making Columbia Investigation board.
Email from Linda Ham to Dittenmore:
From: HAM, LINDA J. (JSC-MA2) (NASA)
Sent: Tuesday, January 21, 2003 11:14 AM
To: DITTEMORE, RONALD D. (JSC-MA) (NASA)
Subject: FW: ET Briefing - STS-112 Foam Loss
You probably can't open the attachment. But, the ET rationale for light for the STS-112 loss of foam was lousy. Rationale states we haven't changed anything, we haven't experienced any 'safety of light damage in 112 lights, risk of loss of bi-pod ramp TPS is same as previous lights...So ET is safe to fly with no added risk Rationale was lousy then and still is....

-----Original Message-----
From: MCCORMACK, DONALD L. (DON) (JSC-MV6) (NASA)
Sent: Tuesday, January 21, 2003 9:45 AM
To: HAM, LINDA J. (JSC-MA2) (NASA)
Subject: FW: ET Briefing - STS-112 Foam Loss
Importance: High
FYI - it kinda says that it will probably be all right
[ORR=Operational Readiness Review, VAB=Vehicle Assembly Building, IFA=In-Flight Anomaly, TPS=Thermal Protection System,
ET=External Tank]
Ham's focus on examining the rationale for continuing to fly after the foam problems.

H.
CAIB Volume 1

MISSED OPPORTUNITY 2 Reviews of light-deck footage confirm that on Flight Day One, Mission Specialist David Brown filmed parts of the External Tank separation with a Sony PD-100 Camcorder, and Payload Commander Mike Anderson photographed it with a Nikon F-5 camera with a 400-millimeter lens. Brown later down linked 35 seconds of this video to the ground as part of his Flight Day One mission summary, but the bipod ramp area had rotated out of view, so no evidence of missing foam was seen when this footage was reviewed during the mission. However, after the Intercenter Photo Working Group caught the debris strike on January 17, ground personnel failed to ask Brown if he had additional footage of External Tank separation. Based on how crews are trained to film External Tank separation, the Board concludes Brown did in fact have more film than the 35 seconds he down linked. Such footage may have confirmed that foam was missing from the bipod ramp area or could have identified other areas of missing foam. Austin's mention of the crew's filming of External Tank separation should have prompted someone at the meeting to ask Brown if he had more External Tank separation film, and if so, to downlink it immediately.

-----Original Message-----
From: SHACK, PAUL E. (JSC-EA42) (NASA)
Sent: Tuesday, January 21, 2003 9:33 AM To: ROCHA, ALAN R. (RODNEY) (JSC-ES2) (NASA); SERIALE-GRUSH, JOYCE M. (JSC-EA) (NASA) Cc: KRAMER, JULIE A. (JSC-EA4) (NASA); MILLER, GLENN J. (JSC-EA) (NASA); RICKMAN, STEVEN L. (JSC-ES3) (NASA); MADDEN, CHRISTOPHER B. (CHRIS) (JSC-ES3) (NASA)
Subject: RE: STS-107 Debris Analysis Team Plans

This reminded me that at the STS-113 FRR the ET Project reported on foam loss from the Bipod Ramp during STS-112. The foam (estimated 4X5X12 inches) impacted the ET Attach Ring and dented an SRB electronics box cover. Their charts stated, "ET TPS foam loss over the life of the Shuttle program has never been a 'Safety of Flight' issue". They were severely wire brushed over this and Brian O'Conner (Associate Administrator for Safety) asked for a hazard assessment for loss of foam. The suspected cause for foam loss is trapped air pockets which expand due to altitude and aero thermal heating

Flight Director Steve Stich discussed the debris strike with Phil Engelauf, a member of the Mission Operations Directorate, after Engelauf returned from the Mission Management Team meeting. As written in a timeline Stich composed after the accident, the conversation included the following. *"Phil said the Space Shuttle Program community is not concerned and that Orbiter Project is analyzing ascent debris...relayed that there had been no direction for MOD to ask DOD for any photography of possible damaged tiles"*

[MOD=Mission Operations Directorate, or Mission Control, DOD=Department of Defense] "No direction for DOD photography" seems to refer to either a previous discussion of photography with Mission managers or an expectation of future activity. Since the inter-agency agreement on imaging support stated that the Flight Dynamics Officer is responsible for initiating such a request, Engelauf's comments demonstrates that an informal chain of command, in which the Mission Operations Directorate figures prominently, was at work.

About an hour later, Calvin Schomburg, a Johnson Space Center engineer with close connections to Shuttle management, sent the following e-mail to other Johnson engineering managers. Shuttle Program managers regarded Schomburg as an expert on the Thermal Protection System. His message downplays the possibility that foam damaged the Thermal Protection System.

However, the Board notes that Schomburg was not an expert on Reinforced Carbon-Carbon (RCC), which initial debris analysis indicated the foam may have struck. Because neither Schomburg nor Shuttle management rigorously differentiated between tiles and RCC panels, the bounds of Schomburg's expertise were never properly qualified or questioned. Seven minutes later, Paul Shack, Manager of the Shuttle Engineering Office, Johnson Engineering Directorate, e-mailed to Rocha and other Johnson engineering managers information on how previous bipod ramp foam losses were handled.

-----Original Message-----
From: SCHOMBURG, CALVIN (JSC-EA) (NASA)
Sent: Tuesday, January 21, 2003 9:26 AM
To: SHACK, PAUL E. (JSC-EA42) (NASA); SERIALE-GRUSH, JOYCE M. (JSC-EA) (NASA); HAMILTON, DAVID A. (DAVE) (JSC-EA) (NASA)
Subject: FW: STS-107 Post-Launch Film Review –
Day 1 FYI-TPS took a hit-should not be a problem-status by end of week. [FYI=For Your Information, TPS=Thermal Protection System]

Jeannette Remak

The approval for launch of STS-51L Challenger by Morton Thiokol's Joe Kilminister.

L.

MTI ASSESSMENT OF TEMPERATURE CONCERN ON SRM-25 (51L) LAUNCH

0 CALCULATIONS SHOW THAT SRM-25 O-RINGS WILL BE 20° COLDER THAN SRM-15 O-RINGS

0 TEMPERATURE DATA NOT CONCLUSIVE ON PREDICTING PRIMARY O-RING BLOW-BY

0 ENGINEERING ASSESSMENT IS THAT:

  0 COLDER O-RINGS WILL HAVE INCREASED EFFECTIVE DUROMETER ("HARDER")

  0 "HARDER" O-RINGS WILL TAKE LONGER TO "SEAT"

    0 MORE GAS MAY PASS PRIMARY O-RING BEFORE THE PRIMARY SEAL SEATS (RELATIVE TO SRM-15)

      0 DEMONSTRATED SEALING THRESHOLD IS 3 TIMES GREATER THAN 0.038" EROSION EXPERIENCED ON SRM-15

  0 IF THE PRIMARY SEAL DOES NOT SEAT, THE SECONDARY SEAL WILL SEAT

    0 PRESSURE WILL GET TO SECONDARY SEAL BEFORE THE METAL PARTS ROTATE

      0 O-RING PRESSURE LEAK CHECK PLACES SECONDARY SEAL IN OUTBOARD POSITION WHICH MINIMIZES SEALING TIME

0 MTI RECOMMENDS STS-51L LAUNCH PROCEED ON 28 JANUARY 1986

  0 SRM-25 WILL NOT BE SIGNIFICANTLY DIFFERENT FROM SRM-15

JOE C. KILMINSTER, VICE PRESIDENT
SPACE BOOSTER PROGRAMS

MORTON THIOKOL, INC.
Wasatch Division

[Ref. 2/26-6]

352

Index

6

651 Requirement, 52

A

Aerospace plane, 51, 52

Air Force, 2, 8, 22, 41, 44, 47, 51, 74, 92, 102, 118, 119, 121, 133, 134, 140, 151, 158, 168, 202, 225, 272, 280, 285, 291, 298, 340

Air Research and Development, 5, 9

Alan Shepard, 25

Apollo, 5, 7, 1, 2, 5, 7, 28, 29, 31, 56, 58, 72, 73, 82, 83, 86, 88, 89, 91, 94, 95, 96, 97, 98, 99, 100, 101, 103, 104, 124, 126, 130, 186, 187, 223, 225, 231, 246, 290, 300, 305, 310, 318, 319, 320, 322, 324, 325, 328, 329, 332, 333, 334

Apollo 1, 7, 94, 95, 231

AQUATONE, 8, 10

ARDC, 5, 9, 10

C

Cape Canaveral, 21, 81, 82, 83, 84, 86, 87, 99, 120, 134, 154, 166, 292

CIA, 6, 8, 10, 11, 12

Cold War, 15, 24, 31, 32, 334

Colonel Curtis Scoville, 41

CORONA, 10, 11

Cuban Missile Crisis, 24

D

DARPA, 18

Defense Advanced Research Projects Agency, 18

Department of Defense, 2, 9, 18, 49, 52, 56, 68, 72, 178, 227, 336, 340, 350

Donald Douglas, 8

DYNASOAR, 32, 34, 35, 36, 37, 38, 40, 48

DYNASOAR II, 37

E

Edwards AFB, 22, 32, 33, 42, 44, 45, 47, 74, 75, 140, 141, 144, 185

Eisenhower, 5, 6, 7, 8, 10, 11, 15, 16, 17, 18, 20, 21, 24, 31, 229

F

FDL, 39, 40, 43, 44, 45, 47, 54, 56, 63, 65, 78, *Flight Dynamics Laboratory*

G

General E.B. Giller, 41

General OJ (Ozzie) Ritland. Ritland, General Osmond

GENETRIX, 20, *high altitude balloon program*

H

Hugh Dryden. *Deputy NASA administrator*

I

IGY. *International Geophysical Year*

J

J-93 General Electric, 38

James Fletcher, 3, 72, 80, 108, 110, 114, 123

James Killian, 8, 17, 20

James Webb, 29, 31, 83, 88, 90, 93, 100

JFK, 24, 25, 26, 31, *John F. Kennedy*

John F. Kennedy, 24, 94

John Glenn, 25, 28

John Manke, 42, 46, *X-24 Pilot*

K

Keith Glennan. T. Keith Glennan, 82

Konstantin Eduardovich Tsiolkovsky, 13

L

lunar landing, 29

Lyndon Baines Johnson, 18, 100

M

M2-F1, 38, 39

Martin Marietta Corp, 41, 44, 45

Military Industrial Complex, 15

N

NACA, 5, 12, 16, 22, 32, 33, 34, 36, 49, 62, 68

National Advisory Council on Aeronautics, 16

National Defense Education Act of 1958, 17

National Space Act of 1958, 1, 3, 336

North American Aviation, 34, 35, 50, 65, 96, 97, 99

NSC5814, 11

P

Peenemunde, 13, 15

PILOT, 41

Project Apollo, 29, 30

Projects Agency, 10, 18

R

RAND Corp., 20, 38

Redstone Arsenal, 27

Redstone Rocket, 82

Robert McNamara, 26, 320

Ronald Reagan, 6, 225

Russia, 10, 2, 7, 13, 306, 330, 334

S

shuttle management failure, 32

Space Shuttle, 10, 48, 49, 60, 69, 75, 99,
123, 124, 125, 126, 129, 132, 133,
134, 145, 149, 154, 172, 177, 188,
204, 225, 229, 255, 299, 303, 319,
346, 348, 350

Space Transportation System, 49, 224,
226, 233, 255

Sputnik, 7, 15, 16, 17, 18, 24, 26

SV-5 *START*, 39

T

T. Keith Glennan, 82

U

U.S.S.R., 15, 17, 18, 20, 31

United States Air Force, 291

USAF, 7, 2, 5, 8, 9, 10, 11, 21, 26, 32, 33,
34, 35, 38, 39, 40, 41, 42, 43, 44, 45,
47, 48, 49, 50, 51, 52, 54, 55, 56, 57,
61, 63, 64, 65, 68, 69, 70, 72, 74, 78,
81, 82, 84, 87, 88, 89, 90, 92, 93, 106,
114, 118, 119, 121, 122, 128, 131,
133, 134, 135, 136,137, 141, 151, 169,
190, 202, 224, 226, 280, 282, 283,
285, 312, 331

USAF Scientific Advisory Board, 33, 44, 51

USSR, 7, 18, 24, 119, 324

V

Vanguard, 8, 9, 87

Von Braun, 14, 21, 27, 86, 92, 101, 125

W

Werner Von Braun, 14, 21, 27, 81, 86,
105

WS117L, 10, 21

X

X-15, 22, 32, 33, 34, 35, 36, 47, 50, 76,
102, 133, 135, 188, 312, 334

X-15 Committee, 33

X-23 PRIME, 44

X-24A, 38, 40, 41, 42, 43, 44, 45, 47

X-24B, 42, 43, 44, 45, 46, 48

X-24C, 47, 48
XB-70 Valkyrie, 2, 26, 38, 40, 254

Bibliography

Report on Presidential Commission of space shuttle Challenger Accident-June 6, 1986 Columbia Accident Investigation Board- August 2003

NASA Budget FY 1994

NASA Budget FY 2003

NASA Budget FY 2012 ARES First Stage Proposal 2009

On the Frontier- NASA Flight Research at Dryden 1946-1981

Rocket Propulsion Elements-Sutton/Biblarz, John Wiley Publishers,2000

Development of the Space Shuttle-1972-81 T. Smithsonian, 2002

Space Systems Failures-Hartland & Lorenz, 2005 Praxis/Springer Publishing

Space Shuttle Log- Gurney & Forte, AERO Press 1988

Comm Check—M. Cabbage & Wm. Harwood, Free Press, 2004

Challenger Revealed – An Insider's Account of How the Reagan Administration caused the greatest tragedy of the Space Age--Richard C. Cook, Basic Books, 2007

Rocket & Spacecraft Propulsion- Martin J.L. Turner, Springer, 2007

Truth Lies and O-rings—Alan McDonald/ Hanser, Univ. of Florida 2009

Prescription for Disaster- Joseph Trento, Random House,1987

Chariots for Apollo- Pellegrino/Stoff, Dover press, 2007

Sputnik to Space Shuttle- Iain Nicholsen, Merrian Webster Press, 1982

Failure is not an option-Gene Krantz, Simon and Schuster 2001

Challenger-Decision to Launch-Diane Vaughan, University of Chicago press 2016 Enlarged ed.

Selling Peace-The conspiracy that transformed the U.S. Space Program-Jeffrey Manber, Collector's guide Publishing 2010

Space Shuttle –Quest continues-George Torres, Presidio Press.

U.S. Space Launch Vehicle Technology-Viking to the Space Shuttle- J.D. Hunley, University of Florida 2008

Major Malfunction-A true story of politics, greed and the wrong stuff. Malcolm McDonnell, Doubleday Press, 1986

The Hypersonic Revolution—Case Studies in the history of Hypersonic Technology 1961-1986—R Hallion—USAF History Office

Space Race-The Epic Battle between America and the Soviet Union for Dominion of Space—Deborah Cadbury, Harper Perennial, 2007

Sputnik-The Shock of the Century- Paul Dickenson, Walker and Co., 2007Entry Vehicles and Thermal; Protection Systems: Space

Shuttle, Solar Starprobe,Jupiter Galileo Probe—Edited Paul E.Bauer Howard E. Colcott, American Institute of Aeronautics,1983

History of the Kennedy Space Center—K. Liparitito, Orville R. Butler, University Of Florida Press , 2007

Inside NASA-High Technology and Organizational Change in the US Space Program Howard McCurdy, John Hopkins University Press, 2004

Spaceflight and the Myth of Presidential Leadership—Howard McCurdy, Roger Launis, University of Illinois, 1997

NASA PAPERS

Into the Proving Ground: Objectives for Human Exploration near the Moon: Nov 30, 2016—Editor Sarah Loff NASA

A Year after Maiden Voyage, ORION progress Continues—December 4, 2015. Brian Dunbar—NASA

To Slip The Surly Bonds...

The Ins and Outs of NASA's First Launch of the SLS and ORION—
November 27, 2015, Editor--Mark Garcia-- updated August 4, 2017

Final Work Platform installed in Vehicle Assembly Building for NASA's
Space Launch System—Linda Herridge, NASA updated August 4, 2017

ORION Spacecraft Progress Continues with Installation of module to test
propulsion system—Brian Dunbar NASA